RTE GUIDE
DES KEOGH PRESENTS MUSIC FOR MIDDLEBROWS
ON RTE RADIO SEE INSIDE
She Used To Be A Goth, But She's
ALRIGHT NOW
RTE 100
RTE GUIDE
Vol. 3. No. 35. AUGUST 31, 1979. Programmes September 1 — 7. Price 15p. (incl. V.A.T.)
Hear 'Ireland's Choice' with Val Joyce and D.J. Friends
RTE Radio 2, Sunday
BELFAST
GALWAY
DUBLIN
LIMERICK
CORK
WATERFORD
RTE GUIDE
21 YEARS OF CORK RADIO A CELEBRATION FRIDAY RTE RADIO 1
Programme as Broadcast.
24th. December, 1926.
STUDIO CONCERT.
Denmark Street, Dublin.
Time. Duration. Artist. Item. Composer. Publisher.
19.25 News Bulletin.
19.30 18 mins. Miss P. Hoey. Christmas Legends of Ireland.
19.48 13 " Miss Crofton. Mangan and Aubrey de Vere.
20.5½ 60 " St. Mary's Choir Donnybrook, under Canon Wilson. (Puer Nobis Nascitur. Old Carol.
(In dulci jubilo. do.
(Song of London Waits. do.
(The Cherry Tree Carol. do.
(I saw three Ships. do.
(Furry Day Carol. do.
(Cradle Song. Provencal Noel.
(When the Crimson Sun. Swiss Carol.
(Over the Hills. Norwegian Carol.
Talk by Canon Wilson.
21.7 47 " Tom Madden. C. Exshaw. Pat Hayden. W. O'Toole. A. Fitzgerald. H. Thomas. "Santa Claus Calling" Sketch. Unpublished
21.55 Interval Announced.
22.2 6 " Station Orchestra. & Louis O'Brien. Xmas Eve. B. Hollander. Hawkes.
22.10 9 " Chorus. Gloria in Excelsis. Beethoven. Novello.
22.20 4 " William Lemass. An Xmas Dream. A. Holmes. Pigott.
22.26 9 " Chorus & Orchestra. Adeste Fideles. Ancient. Novello.
22.37 5½ " do. The Heavens are telling. Hayden. do.
22.42½ Weather Forecast.
22.43 Xmas Greetings.

SOUNDING IRELAND

CLÍODHNA NÍ ANLUAIN

SOUNDING IRELAND

RADIO AND THE MAKING OF US

RTÉ | 100

NEW ISLAND

SOUNDING IRELAND
First published in 2026 by
New Island Books
Glenshesk House
10 Richview Office Park
Clonskeagh, Dublin D14 V8C4
Republic of Ireland
www.newisland.ie

Print ISBN (hardback): 978-1-83594-065-5
eBook ISBN: 978-1-83594-066-2

British Library Cataloguing in Publication Data. A CIP catalogue record for this book is available from the British Library.

Product safety queries can be addressed to New Island Books at the above postal address or at info@newisland.ie.

Front cover Image: DJ and presenter Michael McNamara, in 1980 on an RTÉ Radio 2 (2 FM) outside broadcast in Co Mayo. Courtesy RTÉ Archives.
'Glanmore Sonnet VII' (also known as 'The Shipping Forecast') by Seamus Heaney courtesy of Faber and Faber.

Set in Neue Haas Grotesk Text, 10pt on 14pt
Typesetting and cover design by Emer Brennan, emerbrennan.com
Index compiled by Fergus Mulligan, publishing.ie
Printed by L&C Printing Group, lcprinting.eu

MIX
Paper | Supporting responsible forestry
FSC
www.fsc.org
FSC® C152082

RTÉ

New Island Books is a member of Publishing Ireland and the Independent Publishers Guild.

10 9 8 7 6 5 4 3 2 1

CONTENTS

1926 2RN begins broadcasts from studio in Little Denmark St, Dublin. Wireless Telegraphy Act.

1928 2RN moves to GPO (General Post Office).

1933 Athlone transmitter officially opened by Taoiseach De Valera.

1936 ‘O’Donnell Abú’ becomes station’s identification signal. Acetate disc recording installed in Dublin studios.

1937 Station renamed as Radio Éireann.

1939 Wartime censorship begins.

1940 Experiments in shortwave broadcasting to connect with diaspora.

1947 RÉ Players and Mobile Recording Unit established.

1949 Tape recording introduced to RÉ studios.

1953 Improved studios open in GPO. First audience research shows 83 per cent of population listens almost daily.

1958 Cork studios open.

1960 Broadcasting Authority Act established statutory corporation called Radio Éireann, anticipating RTÉ television’s arrival.

1961 RTÉ Donnybrook Campus opens with 360-foot mast. VHF/FM network of radio transmitters installed nationally.

1966 1916 commemorated with 34 hours broadcasting. VHF/MW broadcasts begin.

1968 All day radio begins.

1972 Raidió na Gaeltachta goes on air.

1973 Live broadcasts begin from Radio Centre.

1975 Tullamore transmitter replaces Athlone.

1976 Final broadcast from GPO by announcer Ray Lynott.

1979 Radio 2 opens.

1984 RTÉ FM3 classical music and arts station starts, sharing frequency with RnaG.

1987 RTÉ launches Atlantic 252 with Radio Luxembourg.

1988 Radio 2 relaunches as 2FM abandoning most non pop/rock music programmes.

1996 Radio 1 goes 24-hours. Radio cars, mobile studios and roadcasters with satellite transmitter systems introduced. Launch of RTÉ online.

1999 Lyric FM launches in Limerick.

2002 Atlantic 252 closes.

2003 Radio stops broadcasting on shortwave, concentrating on satellite transmission and relay via World Radio Network.

2004 RTÉ begins broadcasting on longwave from Clarkstown transmitter.

2007 RTÉ Gold, Digital Radio News, 2XFM, Choice and Digital Radio Sport launched.

2008 RTÉ Chill, Pulse and Jr Radio begin. Digital Radio Sport closes.

2010 Saorview launched, carrying radio and television choice.

2012 Radio regional studios relocate to Galway, Sligo and Waterford Technical Colleges.

2020 Covid 19 Taoiseach Leo Varadkar broadcast on RTÉ that *This is a St Patrick's Day like no other*. Many presenters broadcast from home.

2026 RTÉ100 marks a century of public service media. RTÉ Listen launched.

AUTHOR'S NOTE

At 7.45 p.m. on 1 January 1926, a crackling voice over the airwaves marked the beginning of public broadcasting in independent Ireland. From that first transmission by 2RN – the precursor to RTÉ Radio – grew a common ground of sounds and voices, heard by gradually increasing numbers of listeners across its first century. Along the years, millions of minutes of content have gone out on air to millions of listeners.

Some programmes were broadcast live, others recorded and then broadcast. Some are now in RTÉ Archives. Others were controversial or became iconic. Some shocked, devastated and angered listeners. Others offered sacred and spiritual succour, taught and opened minds, delighted and excited, had us dancing, poked fun at us, made us laugh. They caught daily life, our ordinary selves, taking one step after another, getting on with the getting on of it all. Where we found our voices, put our foot in it, caught ourselves on. Some of it in Irish, most of it in English.

This material forms the backbone of the book. And although much has been written – and continues to be written – about Irish radio broadcasting policies and systems or finances or personalities (and they all figure a little within the pages), the focus of this book is on what was broadcast and what people heard on RTÉ Radio. A sound record of time passing.

But what enabled the station to become this record, this living repository of ourselves in sound in the first place? The Wireless Telegraphy Act of 1926 meant only stations in Ireland with a licence to broadcast could do so and the only licence the department of posts and telegraphs provided was to its own public station, 2RN. In response, illegal or pirate Irish radio stations mushroomed over the years, providing other options for listeners (although in time RTÉ also offered further choice in its Raidió na Gaeltachta (RnaG), Radio 2 (2FM), lyric fm and its digital stations. The Radio and Television Act of 1988 and the Broadcasting and Wireless Telegraphy Act of 1988 meant other licensed national as well as regional, local or community stations could now officially broadcast alongside RTÉ radio stations and have been doing so since. However, RTÉ Radio is unique as the country's only public-service media on air for the hundred years that coincided with the first century of independent Ireland.

I am fascinated how RTÉ Radio has (and continues to) aurally map stages in the story of us making and being ourselves – as much in the present moment as through retrospective telling and recounting. Radio, in sound, captures how our collective and individual voices and ways of communicating have subtly but noticeably evolved over decades of being on the air. What radio offers hovers between being in the moment and recording something larger and longer lasting beyond that moment. I like to think that what is gathered in this book extends to include some of both.

As a programme-maker, choosing material is an integral part of what I do. For this book, my work has taught me to be as curious about what was broadcast in the hinterland as the highways of the daily schedules. All types of radio show us in the making. With such a breath of material available to me, I needed to set myself perimeters: for clarity for you, the reader, as much as for the impossibility of even representing a fraction of what was broadcast. So, while some of what RTÉ Raidió na Gaeltachta (RnaG), lyric fm and 2FM broadcast feature in the book (and RTÉ digital is referenced in the book's timeline), my primary focus has been on what was broadcast on Radio 1 and its precursors, Radio Éireann and 2RN.

Talking to listeners was key for me to gauge radio's place in our lives. What others generously shared illustrated how our relationship with radio reaches deep into our personal stories: one woman, for example, formed her lifelong passion for classical music when she was a child by sitting under the table on which sat the family radio. Another listener, on heading off to train as a nurse, was given a portable radio with a beautiful leather cover; a blue Sobell transistor set was someone else's first radio. One person loved farming and business programmes because they made her appreciate the variousness of the lives of her neighbours from backgrounds and places different to her own. Cracking up over Saturday's political satire with his flatmates is indelibly linked for another listener with his student days at UCD. If the closing sound of *Morning Ireland* started on the car radio, it meant another listener was late for school. And because someone's mother always had *The Gerry Ryan Show* on, she still knows the number of the Ryan Line!

In no particular order, others recall hearing about the Stardust fire the morning after attending a disco in their local school, the Good Friday Agreement declaration on a bus driver's radio, queuing to pay for milk in the local shop when Finance Minister Brian Lenihan mentioned something about a universal social charge in his budget speech, the breaking news of 9/11, the *Blackrock Boys* documentary being on while waiting in another school carpark to pick up their child, or the mother whose boy lit up whenever 'Don't Cry for Me, Argentina' came on the radio in the 1970s. Indeed, my associations with subjects of chapters in the book play occasional parts in their telling, much as I hope yours, or those of others you know, will come to mind as you read through various chapters. Radio is personal. We listen emotionally.

Hours stolen on holiday to listen to archive programmes for the book told me as much. Afternoons lying in a room darkened by closed shutters and curtains to manage unprecedented summer temperatures were ideal for research. Unexpectedly, in came my mother's voice with a pertinent question: she had been in the audience of a programme broadcast over a quarter of a century ago. I hadn't heard it before, nor knew of her presence in the audience.

Hearing her, though, recalled heading off on another holiday around the same time as she would have asked her question. She had kindly offered my boyfriend and I a lift to the airport. We got stuck in heavy traffic. Distracting us (and doubtless herself too) from the delay, she started in the manner of conductor Proinsias Ó Duinn to gesture instructions to the classical musicians playing on the car radio, an imaginary RTÉ Concert Orchestra suspended in front of us, doing its thing on the turn around Lincoln Place on to Westland Row. It was from the radio she had bought herself at twenty-one that I heard most of my formative listening before I acquired one of my own.

But back to the public context of radio-listening. When 2RN began broadcasting, communication worldwide was expanding rapidly. Communication systems that depended on regular post and, more recently, morse telegraphy were quickly joined by telegraph communications to the Antipodes, transatlantic telephony, radio beacons and compasses, aircraft and shipping communications as well as radio broadcasting. By 1926 there were already 151 European radio broadcasting stations, including those in Great Britain and the North. The United States had 536 station, Canada had fifty-two and Australia had twelve.

With 2026 being the centenary year of RTÉ Radio, it was natural for me to look at what was written and broadcast about the anniversary by other programme-makers and contributors at different significant milestones in its story. What I noted on the fortieth, or even fiftieth anniversaries, was much talk about the challenges of the station inventing itself and the newness of it all as well as discussion about other radio stations available in the country in 1926, including BBC and continental stations. Though they were not specifically aimed at Irish listeners, their existence certainly played a part in progressing the opening of the state's own radio service and the finding of its voice after the recent hard-won independence. One hundred years later, a history has been accumulated.

There is no chronological ladder for you, the reader, to climb. That happens naturally in individual chapters. Many episodes could sit in different chapters too – such is the way programmes overlap in subject and in how they are made. I like to think that how the book is laid out is truthful to the experience of radio-listening itself. It can be read through in order of chapters but stands up to incidental reading too.

How we listen – and make – radio has changed over a hundred years. And while we cannot know how listeners will engage with radio in the future, what is sure is that our passion to communicate, to make sense of each other and the world as we go, will find a way.

Working on this book has been a pleasurable process of discovery and rediscovery for me. I hope spending time with it will be the same for you too.

Clíodhna Ní Anluain
Lúnasa, 2026

STATION SOUNDS

CHAPTER 1

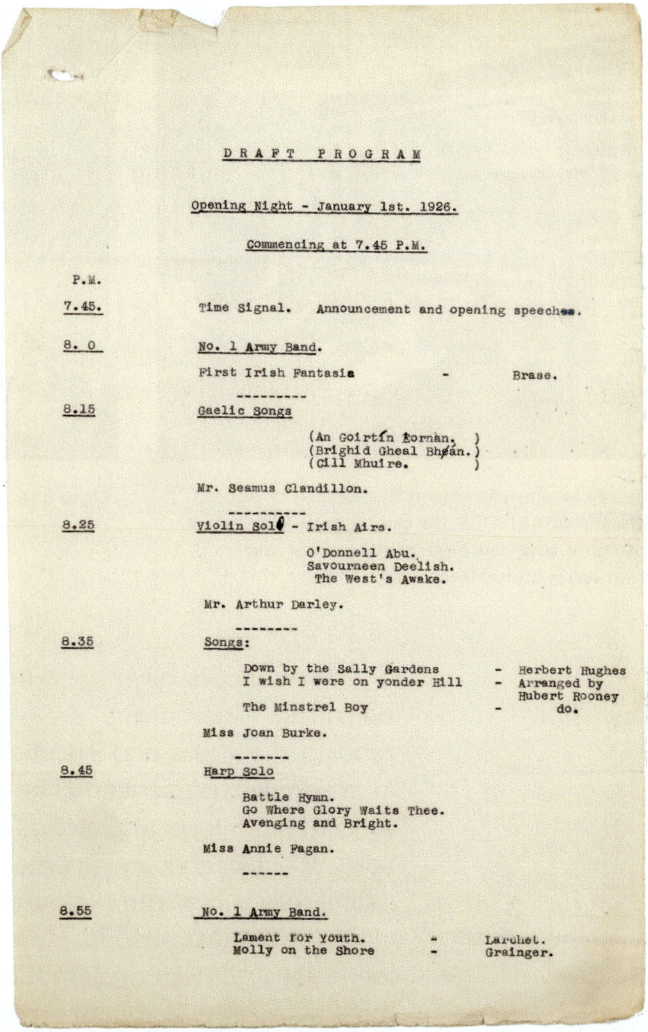

DRAFT PROGRAM

Opening Night - January 1st. 1926.

Commencing at 7.45 P.M.

P.M.			
7.45.	Time Signal. Announcement and opening speeches.		
8. 0	No. 1 Army Band.		
	First Irish Fantasia	-	Brase.
8.15	Gaelic Songs		
	(An Goirtín Eornán,) (Brighid Gheal Bhán.) (Cill Mhuire.)		
	Mr. Seamus Clandillon.		
8.25	Violin Solo - Irish Airs.		
	O'Donnell Abu. Savourneen Deelish. The West's Awake.		
	Mr. Arthur Darley.		
8.35	Songs:		
	Down by the Sally Gardens	-	Herbert Hughes
	I wish I were on yonder Hill	-	Arranged by Hubert Rooney
	The Minstrel Boy	-	do.
	Miss Joan Burke.		
8.45	Harp Solo		
	Battle Hymn. Go Where Glory Waits Thee. Avenging and Bright.		
	Miss Annie Fagan.		
8.55	No. 1 Army Band.		
	Lament for Youth.	-	Larchet.
	Molly on the Shore	-	Grainger.

First of three pages of the opening-night programme of 2RN.

The sound of RTÉ Radio didn't arrive ready-made. Instead, it began to evolve – and has continued to do so – during its first hundred years. As early as 1948, the *Radio Éireann Yearbook* declared: 'Progress in such an institution as Radio Éireann is less often by leaps and bounds than by crawlings and creepings. Short, frequent advances generally pass unnoticed, so that things seem to stand as they were. And there have been such advances.' Some changes happened slowly, subtly. Some have been sudden.

Brigid Kilfeather, Assistant Station Supervisor at Radio Éireann in 1962, the only female supervisor at the time. She was also an announcer and presented Hospital Requests.

Radio Éireann announcer Una Sheehy.

Announcers have been an integral part of radio since the beginning. Their tasks are myriad: introducing programmes, reading the news and weather, notifying delays and changes in schedules, presenting listeners' music-request programmes, reading radio talks and stories, signposting programmes and all sorts of public announcements. They have an unrivalled ability to seamlessly accelerate or slow down talk to a specific time on the clock, to the last second. Indeed, as early as 1 November 1927, journalist J. B. Hall declared in the *Irish Independent* that 'the most compelling and masterful figure in all that building, for the nonce, is unquestionably the official "Announcer". To be a good and true Announcer requires, as I can well believe, qualities in a man as in a poet, that are born in him, not made. He must have an unerring instinct for the job.'

The announcer's work could reasonably be compared to the conductor of an orchestra. It is through their live studio sound desk and microphone that most of what we hear on radio continues to be introduced and put on air. (In much the same way that RTÉ Radio's master control room, deep in the Radio Centre, is where the station's engineers keep the enterprise on air.)

Terry Wogan in RÉ's GPO studio in 1963.

Gay Byrne in RÉ's GPO studio in 1963. He presented *The Gay Byrne Show* from 1973 until 1998, and a jazz, big band and swing music programme on Lyric FM from 2006 to 2019.

Terry Wogan, who became one of BBC's most popular broadcasters, successfully answered a Radio Éireann advertisement for the position of announcer in 1961. On morning duty, he later recalled 'bilingually opening the station having first identified which saint's day it was'. Gay Byrne, who went on to become Ireland's most celebrated twentieth-century broadcaster, started freelancing in Radio Éireann. He remembered how, as a weekend announcer, 'We did Irish language programmes, introduced the orchestras, had our own disc programmes.' As a child I was intrigued by their SOS announcements, something along the lines of, 'Could Mr and Mrs Buckley, currently holidaying in the south-west of the country, and travelling in a green Cortina, registration no. xxxx, please telephone home for an urgent message.' Remember that in the 1970s communication between those on holiday and at home could at best be a John Hinde postcard. I used to wonder what the reason was for these mysterious announcements.

On 8 November 1976 RTÉ Radio made its final broadcast from Henry Street – the station's move to the new Radio Centre in Donnybrook, having started in 1971, was complete.

2RN's first announcer Seamus Hughes was the first person to speak on air on the opening night of the station.

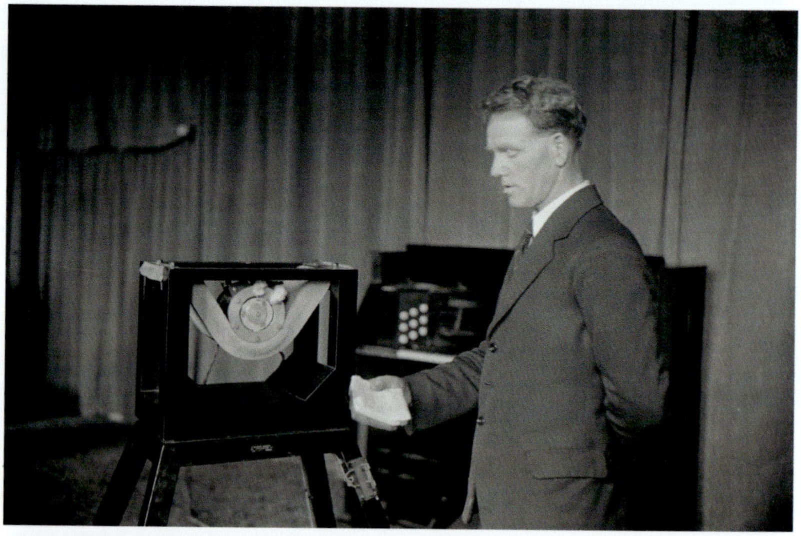

Jan Ní Fhlanagáin, who spoke the first words of RTÉ Radio's second century on air.

Announcers gathered for the farewell with the last message being read by Ray Lynott, his name having been pulled from a hat: 'We say goodbye to you from our Henry Street studios. I think of many people who sat in this chair before me, and I would like to say on behalf of all the present announcers and of myself a very good night to you.'

It articulated the bond between announcer and listeners. RTÉ Radio's other stations, as they were launched, did not have

UACHTARÁN NA hÉIREANN
PRESIDENT OF IRELAND

Message from President Catherine Connolly on the occasion of the 100th anniversary of the first official broadcast of 2RN

1 January 2026

As President may I send my very best wishes to you all as you celebrate this very special anniversary.

The launch of 2RN one hundred years ago on the 1st January, 1926 was a momentous step taken by our fledgling state signifying the start of public broadcasting.

From that beginning a century ago and notwithstanding the name changes, radio became and remains an integral part of all our lives. In and of itself, this is a significant achievement and clearly reflects the intimate bond between the Radio and its listeners.

It is all the more remarkable however given the challenges posed by the vastly changed media and technological landscape that has developed in that period of time.

Indeed, as we move into the second century of Irish public broadcasting, now more than ever, we need an ongoing, open, informed and inclusive discussion on its role in a healthy democracy and a bilingual country.

Today's celebrations however are a joyful reflection on what has been achieved and as President it gives me great pleasure to recognise that achievement and thank you most sincerely for all your hard work.

Guím gach beannacht oraibhse ar fad atá ag éisteacht ar an ócáid suntasach seo. Gabhaim buíochas ó chroí le chuile dhuine a bhí páirteach i gcruthú an chláir speisialta seo agus gach rath oraibh in RTÉ agus sibh i mbun ceiliúrtha sa mbliain amach romhaibh.

Go raibh maith agaibh ar fad.

Catherine Connolly

Catherine Connolly
Uachtarán na hÉireann

Message to RTÉ from An tUachtarán Catherine Connolly on the 100th anniversary of the first official broadcast of 2RN.

announcers, but rather the presenter of one programme hands over the air to the next presenter. When RTÉ Radio marked its first centenary on air at 7.45 p.m. on 1 January 2026, announcer Jan Ní Fhlanagáin read live on air a letter of congratulations to the station from Uachtarán na hÉireann, Catherine Connolly, Ní Fhlanagáin's voice chiming with the first heard on the station one hundred years before, that of announcer Seamus Hughes.

An amusing critique in *The Irish Times* in 1946 illustrates the bias towards so-called 'proper pronunciation' at the time.

RADIO COMMENTARY

Irritating Mispronunciations On Radio Eireann

By Our Radio Correspondent

NONE of the Radio Eireann announcers, readers or actors could give Professor Higgins, of "Pygmalion," as fascinating a series of shocks as the one he experienced when he first heard the speech of Eliza Doolittle; but they could add considerably to his knowledge of queer pronunciations of the English language. One can imagine him, with pencil and notebook, sitting by his wireless set during a news broadcast or a talk, delightedly jotting down phonetic symbols and conjecturing what peculiar combination of birth, education and station in life was responsible for the distortions which he heard.

This does not imply [illegible] our radio speakers, partic[illegible]ur announcers, are bad. [illegible] that some of them, as new announcers, were allowed to read news bulletins before they had had sufficient training, and their stumbling delivery and awkward phrasing [illegible] have given sensitive listeners [illegible]y spasms of irritation. [illegible] even with those imperfections [illegible] were rarely indistinct, and their [illegible]s, were of good quality. Their [illegible] improved rapidly, and one [illegible]n has now reached a high [illegible]d. Indeed, it is a marvel[illegible] high standard when we consider the rare combination of good vocal apparatus, good education, good breeding and good character required to make a good announcer, and when we consider that the B.B.C. pays its announcers on the basis that announcing is a profession, whereas we pay ours as if it were semi-skilled labour.

COMMON DEFECTS

Where Radio Eireann is at fault is in allowing some announcers and actors to persist in minor faults that, by constant repetition, become major irritations. The most common is a shrinking from the broad or moderately broad "a" sound, a defect characteristic of that mincing, effeminate speech known in Dublin as the Rathmines accent and in Belfast as the Malone road accent. Thus, "In *Jenuery bed femine* conditions *hed* been expected in Central Europe, but the *metter hes* been considered *end* the *menner* of relief decided." That is not what was said on the radio, but all those words in italics came in the course of a few sentences prefixed by an announcement that it was the *Reddio* Eireann news. That is the kind of thing that makes me long to shake the speaker by the shoulders and scream: "But that vowel is ah! *AH!! AH!!!*

There is a similar revulsion among some radio speakers from the broad "i." It is particularly obvious in one announcer's reading of the Stock Exchange prices, when we are told the prices of *Arish* Times First Preference, *Arish* Cinemas, Ltd., and Bank of *Areland* shares. These and other mispronounciations are also heard from actors in radio plays, readers in radio digests and readers of reviews and "talks." There is one quite good actor who scarcely ever pronounces "o" correctly. Lately I heard him speaking about a really infernal thing—the *atamic bamb*. There is another person often heard on the radio who always says *fother* for father and *rother* for rather.

The origin of these mispronunciations is not to be found in any traditional way of speaking the Irish or English languages in any part of Ireland. They are miscegnatio. s, and as such should be utterly abhorrent. With regional accents I have no quarrel, and I believe it is possible to have good speech based on any accent—even on the North Antrim or West Cork accents, provided, of course, that due attention is given to articulation.

STANDARD ENGLISH

The question inevitably arises of what we in Ireland should accept as standard English. I do not favour taking Oxford or B.B.C. English as our models. Nor do I favour our own University colleges. In U.C.D. the regional accents, even among the staff, remain unfused. In Trinity the younger men tend either to Rathminize or Oxfordize their accents. The best models, I think, are to be found among the Church of Ireland's older clergymen, whose careers have taken the usual Church of Ireland course—Trinity College, followed by curacies and rectorships in several parts of Ireland (though I would rule out clergymen whose voices have become distorted through an excess of preaching).

I may add that I make this preference as one who is not a member of that Church, and, further, that I hope that the entire team of Radio-Eireann announcers will be seen in front seats at the next General Synod of the Church of Ireland, armed with dispensations if necessary, and enjoying the best available feast of English as it should be spoken in Ireland.

In 1936 a gentle, music-box-sounding rendition of 'O'Donnell Abú' became the identification signal of Radio Áth Luain or Radio Athlone, as the station was known for a while in the 1930s. (It was named after the country's first high-powered transmission mast near Moydrum.) Listeners were invited to select the sound from twenty pieces of music that would be played every time the station was about to go on air. The identification signal would assist listeners with tuning in to the station on hand-dialled radios, distinguishing the station sound from others being broadcast around the same wavelength. Of nearly a thousand votes 'O'Donnell Abú' received 260, 'Fáinne Geal an Lae' received 151 followed by 'The Blackbird' with 88, 'My Lagan Love' with 58, and 'Brian Boru's March' with 52 votes.

Though not required for tuning-in purposes any more, 'O'Donnell Abú' is played daily at 5.30 a.m. to signal the first Radio 1 daytime programme, *Rising Time*, currently presented in Dublin by Shay Byrne through the week and by Lilian Smith in Cork at the weekends.

2RN'S NEW TIME SIGNAL.

The "approximate" Dublin time is no longer being broadcast orally from the Dublin and Cork Stations. Listeners now hear instead the new automatic "six-dot" time signal, which is broadcast every day at 7.30 p.m. and 10.30 p.m. on week days and at 8.30 p.m. and 11 p.m. on Sundays. An extra transmission of the time signal is given at 1.30 p.m. on week days from the Dublin Station only.

The six-dot time signal consists of six short high-pitched musical notes, or "dots," as they are called, sent out at intervals of one second and superposed on the programme. The first "dot" is sent out five seconds before the hour, or half-hour, and the last dot exactly at the hour or half-hour. The first five dots, therefore, serve only as a warning for the approach of he last dot, which is the time signal proper.

The six-dot time signal has been generally recognised internationally as the standard time signal for giving the correct time to broadcast listeners. Whilst it is sufficiently long to enable listeners to check their time-pieces, it is not so long that it interferes seriously with the regular programme.

The time signal apparatus at 2RN is controlled from a standard mean time clock at Dunsink Observatory. It was designed, and the essential parts were actually made, by the Post Office engineers.

***The Irish Times* of 4 July 1929 reports on changes to the 2RN 'time signal', which functioned as a unifying clock for listeners nationwide.**

In his memoir *The Bend for Home*, Dermot Healy remembered how, as a three-year-old in the 1950s, 'You knew dinner-time and breakfast time by the timbre of the voice broadcasting. Walton's brought men in from the fields.' *The Walton's Programme* was a popular music programme, presented by Leo Maguire and sponsored by the music shop on North Frederick Street in Dublin (as he told listeners at the start of every episode). It ended with the words 'if you do feel like singing, do sing an Irish song'. It was one of many popular sponsored programmes.

My RTÉ colleague Mary Elizabeth Bruton had no idea where the phrase ‘Go, go with the Glen Abbey Show’ her parents used to entice her into the car with when she was small came from, until broadcaster Ray D’Arcy mentioned it on air and that he and his wife used the phrase in the same way! Glen Abbey was a hosiery company. The sponsored show featured Frank Kelly, aka Father Jack of *Father Ted*, who also wrote the show’s sketches along with writer and broadcaster David Hanly. The character of Gobnait O’Lunacy from the show is the voice of ‘Christmas Countdown’, the Irish parody of the ‘Twelve Days of Christmas’, a track that continues to be requested by RTÉ listeners during the festive season.

The *RTÉ Guide* of 2 November 1968 introduced major schedule changes and over seventy new programmes. On the cover clockwise from top left: Valerie McGovern, Joe Linane, Townhall Tonight performer, Eamonn Andrews, Hal Roche and Noel Andrews.

The inclusion of sponsored programmes in the schedule was due to the funding structures set out in the Wireless Telegraphy Act, 1926. It confirmed that Irish broadcasting was a state service, administered under the Department of Posts and Telegraphs until 1961. In addition, unlike other public service broadcasters such as the BBC and Radio France, it was financially resourced by three revenue streams: sales licence fees, advertising on air and duties collected from imports of wireless sets and their parts.

Sponsored programmes had a different feel to Radio Éireann live programmes. Their understandably more populist, magazine-like tones (after all, their purpose was to sell a product) leant towards commercially popular music, comedy sketches and drama. Advice of all sorts was breezily offered: everything from household tips to hand care to affairs of the heart. The Fry's Chocolate's programme *The Kennedys of Castleross* was so popular that when Fry's ended its sponsorship, Radio Éireann decided to continue the serial itself.

Over the years the criteria for sponsorship varied from relatively insular, protectionist criteria to more openness to international influences. This was both a reflection of changing government economic policies (the station was after all under government administration) and a response to the fiscal realities of balancing the station's budget. While sponsorship rates were more attractive to Irish-owned companies and those broadcasting in Irish, under the influence of Erskine Childers, who was appointed as minister for posts and telegraphs in 1951, sponsorship was allowed by international companies who had offices in Ireland (once they were not seen to be competing with Irish-made products).

(Top)
The Irish Hospitals' Sweepstakes lottery programme in the 1930s advised listeners abroad to get in touch with their contacts in Ireland about it (because tickets for it were not legally available in many countries).

(Bottom)
Agony aunt Frankie Byrne was the presenter of *The Woman's Page ('Dear Frankie')*, sponsored by Jacob's biscuits. It was the first Irish confessional-style radio show.

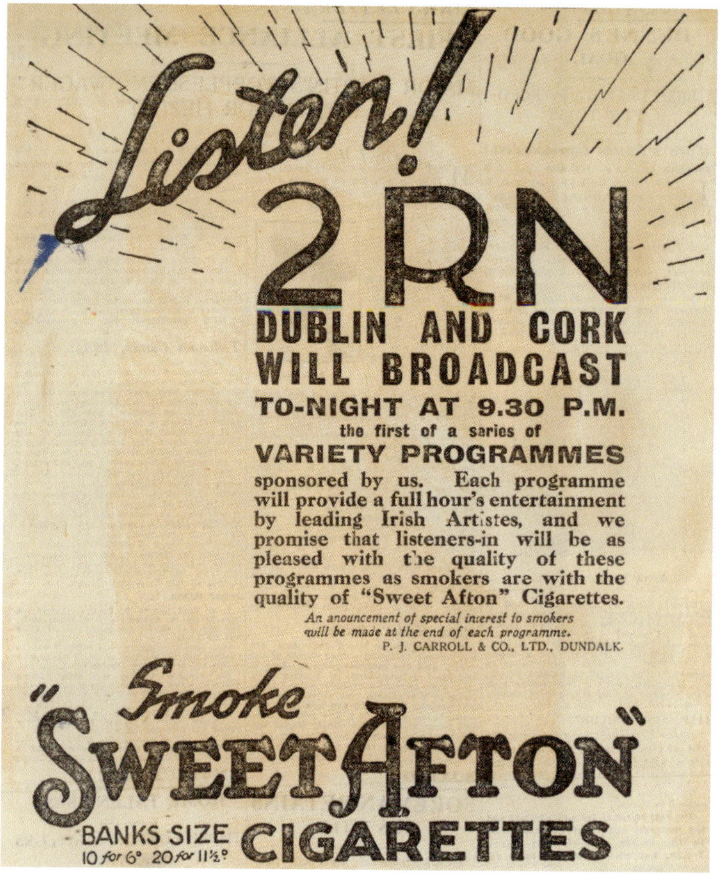

Radio sponsorship came from all sorts of sources. This programme advertised in the *Irish Independent* in 1930 was sponsored by Sweet Afton cigarettes.

November 1968 is a recurring point of change across the station – and features as such throughout this book. It was the same for sponsored programmes when, along with the announcement that morning sponsored programmes were to go, there would be a gradual replacement of them across the day in favour of commercial breaks. Over the following two decades they were gradually phased off the air because RTÉ decided they had become outdated and were determined to update its programming.

The station's first morning programme, *Morning Call*, was presented by Valerie McGovern with Mike Burns and broadcast between 7.30 and 9 a.m. It was RTÉ's first programme that included news, weather, music and features, while short advertisements between and during the programme introduced the commercial breaks we know today.

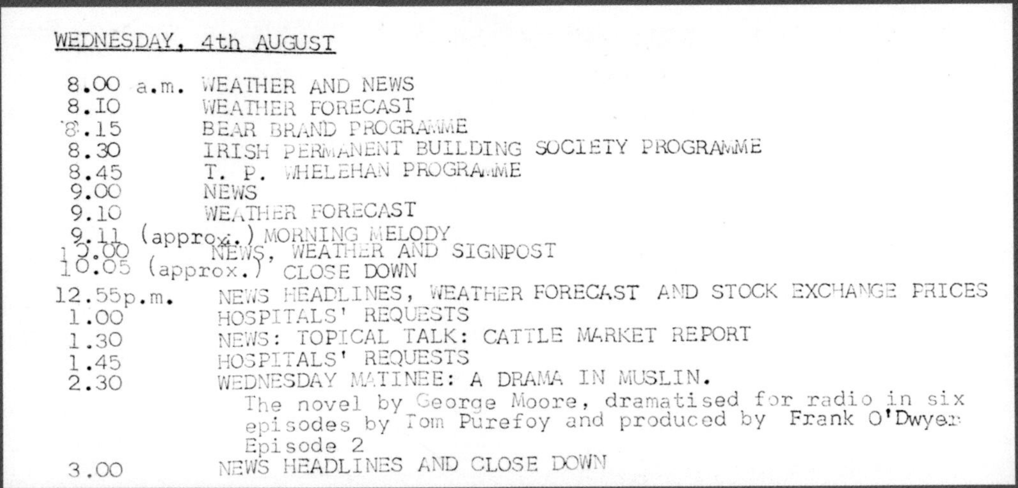

WEDNESDAY, 4th AUGUST

8.00 a.m.		WEATHER AND NEWS
8.10		WEATHER FORECAST
8.15		BEAR BRAND PROGRAMME
8.30		IRISH PERMANENT BUILDING SOCIETY PROGRAMME
8.45		T. P. WHELEHAN PROGRAMME
9.00		NEWS
9.10		WEATHER FORECAST
9.11	(approx.)	MORNING MELODY
10.00		NEWS, WEATHER AND SIGNPOST
10.05	(approx.)	CLOSE DOWN
12.55p.m.		NEWS HEADLINES, WEATHER FORECAST AND STOCK EXCHANGE PRICES
1.00		HOSPITALS' REQUESTS
1.30		NEWS: TOPICAL TALK: CATTLE MARKET REPORT
1.45		HOSPITALS' REQUESTS
2.30		WEDNESDAY MATINEE: A DRAMA IN MUSLIN. The novel by George Moore, dramatised for radio in six episodes by Tom Purefoy and produced by Frank O'Dwyer Episode 2
3.00		NEWS HEADLINES AND CLOSE DOWN

A section of the station's daily report from 4 August 1965 shows the concentration of sponsored programmes in the morning schedule at the time.

Considering that news and current affairs are such a key element of public service broadcasting, their beginnings on the station were inauspicious. Neither featured on the opening night, and a news bulletin was mentioned in the Broadcasting newspaper listings on 24 May 1926 as 'coming from London', but bulletins in the following years were sporadic.

In its first two decades the station failed to generate its own news, and was rather magpie-like, taking bits and pieces mainly from already printed or broadcast sources, including the evening papers and regional journalists. 'There was no newsroom as such, everything was in the general office,' Nancy Bergin, who worked in administration at the station, later recalled. The BBC agreed to allow the station to use its international news for bulletins, and scheduling the lunchtime bulletin for broadcast ten minutes after BBC news meant 2RN news could be confirmed by it before going on air.

However, there were at least two news items that listeners heard in 1927 that originated from the station: one international and one domestic. The station broadcast breaking news when the plane of Charles Lindbergh was sighted flying over Ireland, the first time he was seen since leaving New York on his historic solo flight across the Atlantic. The other was a station announcement which conveyed the minister for posts and telegraphs J. J. Walsh's personal (and as it turned out incorrect) view that Cumann na nGaedheal would not form a government following the result of the general election. Cumann na nGaedheal did in fact go into government, even if it only held power for a few months before another election was called.

Politicians, like the rest of the public, were making sense of the still new medium. Views they expressed about it in the Dáil ranged from criticising the quality of the news, to varying suggestions about whether national or local politics should be reported on at all. One saw political bias in the fact that opposition leader W. T. Cosgrave's winning bull at a local agricultural show did not get a mention, and that if it had been de Valera's, it would have been broadcast. Though politicians grew more interested in radio, it didn't mean they understood it, nor that they yielded more support for the station.

Results of the 1938 general election, which saw de Valera hold on to power, were covered at length by the station, which stayed on air until four in the morning to cover the count. That same month, it broadcast the inauguration of the first president of Ireland, Douglas Hyde.

(Left) ***Here and Now*** **presenter Rodney Rice featured in the** ***RTÉ Guide*** **in 1975. His two-hour weekday late-morning programme was similar to** ***Today with David McCullagh*****.**

(Right) American aviator Charles Lindbergh (1902–1974) with the ***Spirit of St Louis*** **before his epic flight from New York to Paris in May 1927.**

Newsreader Bernadette Plunkett recalled station director T. J. Kiernan's instruction that the declaration of war in 1939 was to be treated calmly. Before the war Frank Gallagher was deputy director of the station. Now he was its censor. But at least the war had hastened the appointment of Michael Lawlor as the station's first news editor in 1945. By the end of 1947 plans were in train to hire additional staff in news. The following years saw reporters recruited along with some 150 news stringers around the country, who would be paid when a story they provided was included in the news. In 1948 two descriptive news writers, more like what are reporters today, were appointed. Brian Durnin and P. P. O'Reilly's work was to write and read news reports and sometimes use recordings made by the mobile recording unit in them to animate a story.

The Cork Examiner began to act as a news agency for the station too, while by 1948 they were also getting foreign news from Reuters. Post-war censorship began to be relaxed around this time too and live broadcasts from abroad began. Trade talks in London between Ireland and England were reported live by Radio Éireann staff from BBC studios, with their up-to-the-minute bulletins bringing an immediacy to the coverage that print media could not hope to imitate.

In 1948 the repatriation from France of William Butler Yeats's body to Drumcliffe, Co. Sligo was a major event for the RTÉ newsroom: P. P. O'Reilly provided background information

The crew of Irish naval vessel *L.É. Macha* on 17 September 1948 before W. B. Yeats's coffin came ashore at Galway to continue its journey from France to Drumcliffe in Sligo.

on Yeats's life and work, sourced an eyewitness account of the departure ceremony from France and was present at the landing scene at Galway docks and the arrival of the cortege into Sligo. With only an hour to spare after the religious service and talk with the Rector of Drumcliffe he returned to Sligo Post Office to prepare his report in time to deliver it live on the 6.30 p.m. bulletin.

A national newspaper strike in 1952 provided the opportunity for Radio Éireann to step up its news coverage. Denis Brennan presented a daily news bulletin for the duration of the strike. It proved welcome to listeners and continued after the strike was over.

News and coverage of political affairs began to settle, preparing the groundwork for their general transmission in future decades. In 1953 *News and Nuacht* headlines began to be broadcast at the hour or half hour, as they are today, so that listeners could more reasonably catch them. A short report of the day called 'Today in the Dáil' was introduced for when the Dáil was sitting. And a year later *Topical Talk*, which followed the

The panel of *Round Table on World Affairs*, chaired by Jack White, furthest right.

2 RTV Guide, December 8, 1961

Closing an old door — and opening the new

Radio Eireann's News Department will be oving this week-end to the new Television entre at Montrose, on the Stillorgan Road. his is part of the plan for streamlining RE's nd Telefis Eireann's news services which will eep you up to the minute on Irish and world ents. Here JOHN ROSS, RE newsman, recalls me colourful memories of ten years in the ld Newsroom in Henry Street.

The " comfortable disorder " John Ross remembers is reflected in this scene in the old R.E. Newsroom in Henry Street. Typewriters, telephones everywhere, but News Editor, Michael Lawlor (centre), and colleagues Seamus O Morain (left) and Mick Hanley contrive from all the seeming chaos the regular news bulletins by which you can set your clock.

Colleagues of news staff member John Ross in their GPO office before their move in 1961 to the new Telefís Éireann campus and a news department that would jointly serve radio and television.

lunchtime news, featured a more detailed report on one particular story of the day.

Round Table on World Affairs chaired by Jack White, which began in 1954, was an example of live political debate in its infancy. There was a nervousness about what might be said if participants were allowed to broadcast live, so programmes were recorded on one day and broadcast the next. But such caution would come back to bite the producers. In September 1959 Radio Éireann's Jack White chaired a discussion on French policy in Algeria. Between the recording of the programme and its broadcast, General de Gaulle declared France's intention to withdraw from Algeria. The broadcast went out with no reference to the general's decision. The station's cautious policy on live political debate was clearly untenable, and soon after current topics were limited to live broadcasts only.

Before the establishment of Comhairle Radio Éireann by Minister for Posts and Telegraphs Erskine Childers in 1953, there had been little or no political debate on the airwaves. Childers was keen to introduce party-political broadcasts such as the ones aired by the BBC. Musing about their introduction, Maurice Gorham, the next director of the station, remarked in 1966 how 'broadcasting people are constantly running themselves into difficulties and complications that they have no need to incur.

There was no necessity for Radio Éireann to put on general election broadcasts at all. It insisted on doing them, and they caused more trouble than many series with much greater programme value. But it is only by tackling this sort of enterprise that a national broadcasting service can feel it is really doing something.'

In anticipation of the arrival of television, an RTÉ news division was established in 1959 to serve both media services. In time it spawned a legacy of current affairs programming and concepts that continue to this day: *World Report*, *This Week*, 'It Says in the Papers' and *Morning Ireland*, not to mention its provision of the station's news bulletins. These programmes have brought focus to political and topical current affairs – from national crisis to budgets, to general and international elections – carving out key areas of listening to Irish current affairs and world affairs from an Irish perspective.

Sean O'Rourke presenting the Radio 1 2007 election results coverage. Though there are several panellists, including former taoiseach Garret FitzGerald (back to camera in black jacket), there is a notable lack of women present.

'We're going to take a step back to 1979' was how presenter Marian Finucane moved on to another subject on her radio programme in March 2015. It was International Women's Day. She set the scene: 'At the end of May 1979, at the same time as RTÉ Radio 2, *Women Today* was launched. Contraceptives were still outlawed. Women were barred from many pubs and very often from having pints.

▲ (Left)
Women Today presented by Marian Finucane featured in the *RTÉ Guide* of 6 July 1979.

(Right)
By December 1986 *Liveline* presented by Marian Finucane was gathering momentum.

Clare Duignan: she and Betty Purcell produce the RTE Radio 1 morning programme.

Producer Clare Duignan, who proposed a women's programme in 1979 following the results of an RTÉ survey, which highlighted the under-representation of women on the air.

Women presenters and producers of RTÉ Radio's features and current affairs department with their department head Donncha Ó Dúlaing in March 1969. Front row from left: Annette Cusack, Clare Boylan, Nodlaig McCarthy. Back row from left: Máire Ní Mhurchú, Christina Murphy, Deirdre Ní Laoghaire, Pamela Perry, Terry Prone, Kathrine Rose and Áine McEvoy.

Marital rape was still legal. Women couldn't have a mortgage or a credit card, were not the recipients of their children's allowance.'

Women Today was one of a series of radio programmes that would include *The Gay Byrne Show*, *Liveline* and *The Gerry Ryan Show* that embodied that feminist slogan, 'The personal is political.' The shows' live phone-in format allowed them to break new ground in Irish broadcasting, where people could talk live on air about what mattered to them, calling from their own phone or from a public telephone. They didn't have a built-in delay. There was no filter. Listeners heard the direct experience, from all sides, of what were called 'unmarried mothers', of marriage breakdown and of alcoholism that shattered families. You could learn what was really going on in Ireland behind closed doors. Former RTÉ producer Clare Duignan remembers that Finucane – who hosted both *Women Today* and its successor on television, *The Women's Programme,* as well as *Liveline* – had been described as the flea in the marriage bed of Ireland by a priest in Maynooth.

For Doireann Ní Bhriain, a reporter on *The Women's Programme*, and its presenter after Marian Finucane, her 'strongest memory is of reading and writing letters ... putting women in touch with each other. Providing information they might otherwise not easily find.'

Of course, women had been working in the station and on air from the beginning. There was trailblazer Máiréad Ní Ghráda

(Left) An all-female audience and (right) panellists during a live *Women Today* broadcast June 1981 about the general election. From left: Fine Gael politician Gemma Hussey, Senator Mary Robinson (later President of Ireland) and Fianna Fáil TD Eileen Lemass.

in her role as first Woman Organiser and Relief Announcer, although the voices she introduced were generally those of men. Kathleen Beatty, known as Kitty Roddy, took over Ní Ghráda's role after 1935. John Brennan – the pseudonym of Sydney Gifford, one of the revolutionary Gifford sisters – covered many subjects, as did Hanna Sheehy-Skeffington. Maud Gonne MacBride was interviewed and Teresa Deevy's plays were broadcast. Kate O'Brien was a book reviewer. Geraldine Neeson's piano-playing, as station accompanist, was heard on the short-lived early Cork iteration of the station in the late 1920s, along with many other musicians and singers who were women.

A weekly women-oriented magazine programme was a Radio Éireann staple of the middle decades of the twentieth century. *The Women's Club*, presided over by Fay Sargent, covered 'all aspects of interest to women'. Among the subjects were cookery, fashion, health, beauty and careers. *Between Ourselves*, produced by Petronella O'Flanagan, had Kate O'Brien on regularly and *Scrapbook for Women* was a musical and literary half hour. *Woman's Magazine*, presented by Oonagh Wirther, was 'for the busy housewife, the not-so-busy bachelor girl, even the harassed husband'. Listeners were invited to get in touch with ideas, suggestions, hints, recipes and items that might be of interest to housewives. The programme included reports from the continent; interviews with exterior and interior decorators;

Acclaimed Limerick-born writer Kate O'Brien was a regular contributor to RTÉ Radio and its precursors. Her once banned novel *Land of Spices* was serialised and read by Joan Sheehy in 1999.

Maire Ní Mhurchú was appointed Head of RTÉ's Cork radio studios in 1977. Her documentaries included *The Echo Boys*, about the boys who sold the Cork newspaper; *Someone to Love* on life in orphanages in the 1960s; and *Children Talking*.

Kathleen Dolan presenting Radio Éireann's *Hospitals' Requests* programme in 1948. According to the Radio Éireann Annual Report, 'she has handled the programme for some years, with conspicuous success'.

Miss Eileen Ware, a lady architect on the layout of kitchens; the principal of a school of dress-designing talking about saving money by making your own clothes; slimming without dieting; exercise; and the houses of Copenhagen. Some of the subjects were certainly interesting and akin to those broadcast on similar programmes today. However, the fact that the women and the programmes can be so easily captured in a paragraph or two highlights the relative absence of women on air compared to men during the first half century of the station.

In the November 1968 expanded radio schedule, there were signs of change in the presence of women. Valerie McGovern was made co-presenter (with Mike Burns and then Sean Duignan) of *Morning Call*. Freda McGough was given *Day to Day*, which started the trend of current and topical affairs being covered live in magazine format. (It was an early forerunner of the present *Today Show*). Nevertheless, Valerie McGovern (who had broadcasting experience in BBC) remarked in the May 1973 *RTÉ Guide* that most people regarded her as a 'wallpaper music type presenter', and that it was ridiculous how rarely women presented features, news programmes or quizzes. 'It would be very satisfying to be able to tell a man he was wrong in an answer to a question!'

The Gay Byrne Show had begun a couple of months before McGovern's interview, gradually opening the doors to women's stories being aired. The programme had an unerring feel for what made good radio at the time and was usually sympathetic to female listeners and their needs. It had nearly an all-female production team, though the male clout brought by Gay Byrne and, producers Bill Wall and in particular John Caden cannot be underestimated in 1970s and 1980s Ireland.

The Gay Byrne Show producer Lorelei Harris emerged from the programme with ambitions to make innovative programmes of her own. And she did. Her Prix Italia award-winning 1995 documentary Dreaming of Fat Men was about women and their relationship with food and their bodies. Critical to the future sounds and voices of RTÉ Radio was that she used the influence of her award to build the internationally acclaimed Doc on One series, a source of thousands of radio documentaries on multiple subjects, including gender, identity and representations.

***RTÉ Guide* of 25 May 1973 featuring Valerie McGovern.**

(Opposite page) John O'Donohue, presenter of *Involvement*, made the cover of the *RTÉ Guide* in 1973.

(Opposite page) 'Talk to Joe': Joe Duffy on *Liveline*. It was on his programme that news of 9/11 broke, and they quickly linked in live with journalist Conor O'Clery in New York.

Programmes such as *Women Today* and *The Gay Byrne Show* can be viewed as part of a broader change within the station's relationship with the country from the sixties, as the public found its voice in talking out loud and the station learned to listen. In the early decades of 2RN, there was a much greater focus on scripted radio, and the station had a certain didacticism to it. Rarely were listeners themselves heard from. There were myriad reasons for this, one of which was the lack of infrastructure – it would be much later before most homes had a telephone. Another was that the nation, still coming to terms with itself in 1926, took some time to gain the self-confidence, and confidence with the broadcast medium, to speak out on air.

As early as the mid-1950s, a well-meaning radio show that asked for listeners' views, such as *What's on Your Mind?*, didn't always work. Its producer, Maxwell Sweeney, would later remark: 'People hadn't yet figured out that radio was something in which they could have a part. Long before the concept of the phone in when everyone talks their head off, very difficult to get them to go!

But by the 1970s, much had changed. Alongside the programmes already mentioned, *Involvement* was another radio programme that welcomed listeners to contribute directly on air. Broadcast on Friday evenings, with the phone lines open from early on in the evening, it hosted wide-ranging debates on controversial topics, from the transmission of the Angelus to control of natural resources, and audience participation in the form of questions and contributions was encouraged. Sometimes these contributions could lead in unexpected directions: one young girl spoke in hushed tones about the problems she had in her home as her parents had just come in from the pub ...

Through the 1970s the view that local and personal voices deserved to be shared was taking root. Community and regional development and its possibilities were a growing international movement. Radio 1's next move in this direction was *Liveline*, established in 1985 with Marian Finucane at the helm. Producer Hilary Orpen in its first month described it as 'a programme which looks beyond the scenes at people, their situations, preoccupations and efforts to make a better life in Ireland'. (In 1985 the country's rate of unemployment was at 17 per cent). She mentioned the divide between rural and urban life and wanting to bring awareness of both to the population, and that *Liveline* was for both men and women.

The programme initially leaned on 'experts' with a panel discussion on Fridays around a topic from that week's calls. Regional reporters also contributed, demonstrating that it was for the whole country. These structures were soon abandoned as callers made the programme their own and this was wholly embraced by *Liveline* as its signature voice. Joe Duffy became presenter in 1999, a tenure that grew the catchphrase 'Talk to Joe' (and 'wash your hands, wash your hands' during Covid). Subjects that carried over several programmes, many of which made media headlines, included the collapse of the banks and their impact on people's savings and lives more generally, institutional abuse, racism, and the menopause. Kieran Cuddihy took over the role in 2025.

Presenter, producer and guests in studio just before the live broadcast of *Involvement* in June 1972. A discussion on the Irish language took place on this edition of the programme.

Gerry Ryan was one of several DJs recruited by RTÉ for its new station Radio 2 (later RTÉ 2FM) on its launch in May 1979. On his late-night programme *Night Lights*, he began to experiment with taking phone calls and what it could be like to engage directly with listeners, in the relative freedom of late-night radio. The previous year he had caught public attention for his outspokenness and capacity to shock. *The Gay Byrne Show* had challenged him and some others to live in the wilderness of Connemara for a week using an SAS survival handbook. He quickly established himself as the spokesperson when they talked to Gay Byrne every morning.

Ryan described how he and his fellow participants had supposedly beaten to death a newborn lamb to cook and eat it – the lamb had in fact been shot for them by a farmer. The nationwide denunciations and demands that the participants should be prosecuted for animal cruelty only brought further attention to Gerry Ryan. His ability to attract the spotlight, as well as what he was already doing on his 2FM programme, landed him a gig as presenter of *The Gerry Ryan Show* for twenty-two years, from 1988 until his death in 2010. The show featured for the first time on the cover of the *RTÉ Guide* of 26 August 1988.

Broadcasters Brendan Balfe (left) and Pat Kenny at the launch of Radio 2 (now 2FM) in 1979. Brendan Balfe's was the first voice heard on Radio 2. Pat Kenny had his own music programme on the station called *The Outside Track* while continuing to broadcast on Radio 1.

'I call it "Risks Incorporated" because most of the time we sail rather close to the wind,' he said about it in his interview with Linda Kavanagh: 'The emphasis is on craic, but there's also a large section that deals with consumer and personal problems. A mix. The programme is 'complete mayhem – the door to the studio never closes. It's rather like Grand Central Station, or a doctor's waiting room. In America, they call it "Zoo Radio"!'

Ryan's programme introduced a new style of programming not only to Radio 2, but to the country: 'It was a risk but one that seems to have paid off ... it was felt it was time for a change, but we weren't quite sure what format to use ... it was decided to work out a basic format and let the listener decide from there on ...'

Ryan could claim himself to have been the first Irish voice blatantly determined to attract the younger radio listener to the national airwaves. I later heard friends I would not associate with phoning a radio show almost confessionally admitting that they spoke on air with Gerry Ryan, sharing things they would never have spoken out loud to others about, from mishaps on holidays to having an abortion, to their battle with their identity. It was the *Gerry Ryan Show* Lavinia Kerwick called in 1992 following the court case that resulted in the conviction of William Conry of her rape. Her public declaration of who she was, when the legal system had protected her identity, had a game-changing effect on countless listeners that morning and subsequently much wider national awakening for what women endure.

Indeed, a generation and more, well in their middle years themselves, came of age listening to him, relishing his bravado, his winding up the official line wherever he could, talking about intimacy and the body, not least his own. A listener told me that when she worked in a clothing factory in Dún Laoghaire the floor staff came to the arrangement that they gave one hour to Gay Byrne and the other to Gerry Ryan, rotating the privilege of who got the first hour, from day to day.

(Left)
Gerry Ryan in 1988, soon after he started presenting his own show on Radio 2 (now RTÉ 2FM).

(Right)
For Valentine's Day 1989 the show called on listeners to tell them 'about their flops and disasters in the Valentine stakes. Being insincere, unkind and unfair, we love getting a good laugh out of other people's misery.'

Brenda Donohue, *Gerry Ryan Show* reporter, on a mobile phone in The Square shopping centre in Tallaght, Co. Dublin in 1993. She was there to promote the annual 2FM/Lions Club Christmas Food Appeal.

Of course, much as what was talked about and how we talked with one another on radio reflected our evolving selves, so too did the broadening out of Irish and international music broadcast by 2RN and on to RTÉ Radio. The uproar about jazz and its unofficial banning for a period in the 1930s, the popularity of commercial releases of stars like John McCormack, Margaret Bourke Sheridan and Delia Murphy, the enormous popularity of céilí bands, the showband craze, the establishment of a charts show, country and western, the blues and rock, alongside traditional and classical music, catered for from the beginning, chart an increasingly more varied response by RTÉ's radio stations to what listeners wanted to hear.

The first hundred years of broadcasting by RTÉ Radio has been one of both continuity and change. Some elements of the 'sound' of Radio 1 in particular have a long history – from continuity announcers, and news and weather reports, to the familiar notes of 'O'Donnell Abú' that signifies the station coming on air in the morning. But it has also been a century of change. The opening of the airwaves to the people of Ireland to speak about their own lives has been a change that has made Irish broadcasting immeasurably richer, and has opened up the country to a conversation with itself.

RTÉ BUILDINGS

SPOTLIGHT

Two buildings, both designed for cutting-edge communications of their day, are synonymous with RTÉ Radio's first hundred years. The first – in Francis Johnston's nineteenth-century Dublin General Post Office design – delivered an iconic building worthy of a great capital city and a thriving British imperial postal service. Later, Scott Tallon Walker's twentieth-century Radio Centre provided a purpose-built home for independent Ireland's public radio service.

Most of RTÉ Radio's broadcasts from its first half-century came from studios converted from the department of posts and telegraphs' offices on the upstairs floors of the GPO, a building and government department inherited on independence. It was from there that the station gradually expanded its archives, disc collection, orchestras, variety of programming and broadcasting hours. It was the place where many of the people mentioned in other chapters conceived and broadcast their programmes; and the place to which RTÉ Radio returned on 15 January 2026 for a

day-long salute to a century on air, acknowledging the physical and imaginative connection that the building still holds for the station and its listeners.

Most of RTÉ Radio's broadcasts during the second half of its first century came from the Radio Centre. Scott Tallon Walker also designed the RTÉ Television Centre and other key buildings that formed part of the wider modernisation underway in Ireland at the time. The Radio Centre is a compact broadcasting village; a circular complex within a rectangular-shaped building, where the rhythm of movement, staff and contributors going to and fro between desks, studios and the sound library, together with their vocal levels and urgency of speech, reveal the exact time or day of the week from the programmes being prepared or broadcast. And then there's the orchestra, whom you might encounter at full flow as they rehearse in Studio 1.

The *RTÉ Guide* marking the fiftieth anniversary of RTÉ Radio, emphasising the modern equipment it uses in the Radio Centre.

(Opposite page) The Radio Centre at night, soon after it was fully operational.

Buildings, and what goes on inside them, can evolve and change. RTÉ Radio provides an example of this. No longer operating in a single building, it now also broadcasts from other more recently built studios elsewhere on the RTÉ campus, and has long had studios around the country, including at Cork, Galway and Limerick.

In 1979 RnaG established its Dublin base in the building. That same year, the Radio Centre welcomed a new station, Radio 2 (now 2FM). 2FM's musical heritage has become synonymous with Studio 8, which has played host to many iconic musical artists, including Oasis, Sinéad O'Connor and The Cranberries. Other musical acts performing on programmes as various as *The Rolling Wave*, *Oliver Callan* and *The Louise Duffy Show* have picked up the musical baton passed on by earlier programmes made in the same iconic studio since the 1970s. Lyric fm, though with its headquarters in Limerick, also broadcasts daily from the Dublin Radio Centre.

If names such as Malin, Mizen and Loop Head conjure up the sea-area forecast, then Kippure, Athlone, Tullamore, Mount Leinster, Mullaghanish, Maghera and Trusk are redolent of the country's transmitter stations. It was through them, on medium wave, FM and VHS transmission systems, that Irish programmes were broadcast well before the introduction of satellite and digital transmission. These transmitter stations are reminders of the physical infrastructure needed to carry sound, even though sound itself respects no physical borders in reaching listeners.

(Left)
Aslan performed on Radio 1's *Today with Pat Kenny* on 21 May 2012 in Studio 8 in the Radio Centre.

(Right)
RTÉ's orchestra, choirs and musical groups have recorded film soundtracks, live broadcasts and rehearsals with several world-renowned artists in Studio 1.

Slip entrance for the transport of musical instruments and equipment in and out of Studio 1 seen during the construction of the Radio Centre.

RTÉ displays information on the newly constructed Radio Centre in the RDS, Dublin, in 1973.

Transition year students on work experience in the Radio Centre in 2017.

AFTER THE BALL

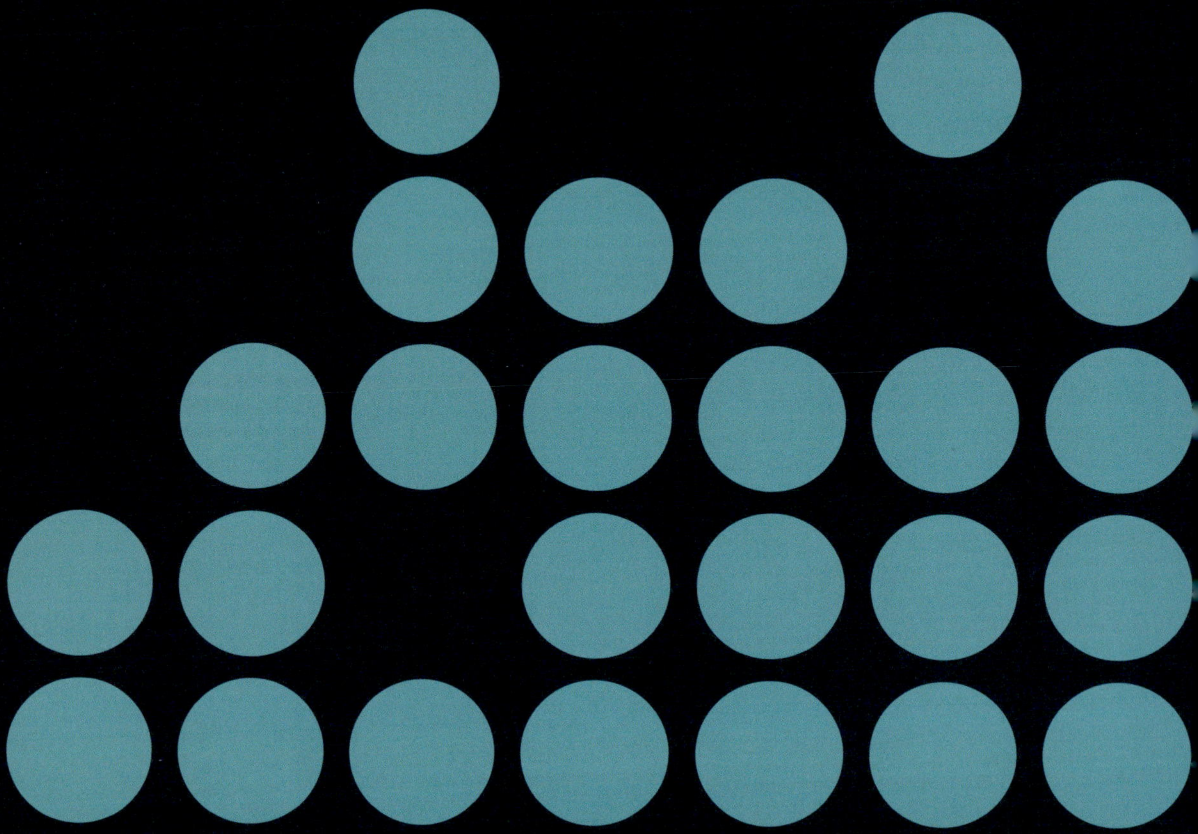

CHAPTER 2

RTÉ Sport's Siobhán Madigan interviews Katie Taylor for 2FM and the Sports and News bulletins following her Once Upon a Time victory over Flora Pili in Croke Park on 5 September 2026.

Each of us who loves sport has their own live radio memories. Where we were, who we were with, not to mention the outcome of contests, how they landed. John McGahern associated the radio with his father rather than his mother. He recalled there was a radio in the living quarters of the barracks where he worked and how on All-Ireland day it would be full of people listening to Michael O'Hehir's broadcast. Commentator Éamonn de Barra treasured a letter sent to him by former Cork captain Seán Óg Murphy. Murphy had written to de Barra to explain that, not having been able to attend the match, he had listened to his commentary on the radio and enjoyed it so much that he no longer felt the need to travel to All-Ireland matches – the commentary was so good it brought it to life for him. A woman still remembers the sheer excitement of Louth winning the All-Ireland in 1957, of listening in on the radio at home with the same exhilaration as her brothers who had gone to Croke Park and witnessed the match at first hand. As recently as November 2024 a letter to *The Irish Times* expressed that, 'further to recent letters concerning excessive chatter by our enthusiastic sports commentators, while less is more, I believe for radio you need the added colour of those wonderful, meandering, descriptive sentences that transport you pitch-side with your boots on, Mícheál Ó Muircheartaigh-style'.

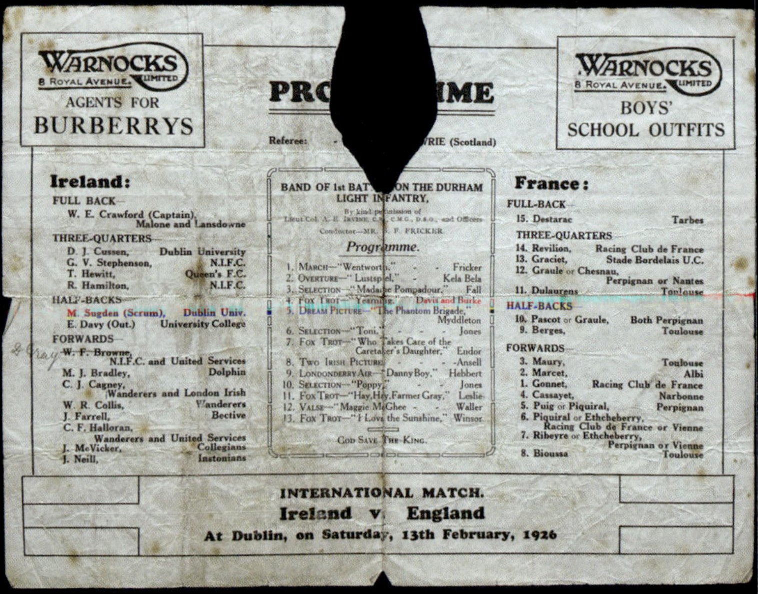

(Left) The cover of the programme for the Rugby International match between Ireland and France played on 23 January 1926 in Ravenhill, in Belfast.

(Right) The inside page of the programme showing the names of the Irish and French players. Ireland won 11–0. The team had been announced earlier in the month, the first reference to sport on the new Dublin station 2RN.

Sport has long served as a conduit for a wide range of subjects and ideas in radio-listening. It can be set off by a conversation with a sporting legend, the commemoration of an important year for a particular sport or discussions linking business and sport. It may also take the form of analysis, phone-ins on current controversies, documentaries, quizzes or programmes where sport is the central theme – whether in a talk, a show or a book inspired by it. Even the brief sports bulletins contribute to this landscape.

These different forms of sport-related listening did not all appear at once but developed gradually as radio evolved, although some have been present since broadcasting's earliest days. Nestled amid the nightly programming of the first month of music and short talks, language lessons, weather reports and SOS notices was the Irish rugby team line-up to play France at Ravenhill on 23 January 1926. They are significant for being the first sports people to have been named on Irish radio. Among the sport mentioned on the station during its first summer were the results of a Garda boxing contest on 17 July and of an Aonach an Gharda athletics event in Croke Park; and a bulletin from the Dublin Horse Show in the RDS on 5 August. Clare Cooke gave a fourteen-minute talk on her impressions of the Horse Show on 6 August. On 22 August results were broadcast of a football match between Kilkenny and Galway and a hurling match in Dungarvan, Co. Waterford.

Live sports commentary began on radio and may indeed be one of the reasons why we indelibly link the voices of commentators with our experiences of sport, with people mentioning the commentator as much as the sport in question when recalling radio moments: to name the commentator is to name the code. Indeed, the point was illustrated in the 2024 RTÉ series and podcast *Chapters of Magnificence*, produced to mark the centenary of Ireland's first participation in the Olympics following Independence. Each episode focused on one Olympian, and mentions how the live commentary of their performances, even for them, has become integral to their memory. After presenter Greg Allen played back the commentary of the last seconds of his race for John Treacy (marathon silver medallist in the 1984 Los Angeles Olympics), Treacy said it was lovely to hear Jimmy Magee's voice. He said when he met people after the Olympics the first thing they would do was congratulate him on his medal and the next thing they would do was to say, 'Jimmy Magee's commentary, wasn't it fantastic!' With Magee telling us Treacy had a hundred metres to go, the commentator went on to name, in order of bronze, gold and silver, all the previous Irish Olympic medallists, perfectly timing his commentary as John Treacy crossed the line to finish with, 'and for the thirteenth time an Irish medal goes to John Treacy!'

Among the outside broadcasts during the first months of the new station were GAA match commentaries from Croke Park, the Cork Athletic Grounds and the Athletic Grounds Thurles, as well as two Irish Rugby Football Union matches from Lansdowne Road, Dublin. At the time Paddy Mehigan was the GAA correspondent for *The Irish Times* and had a weekly article in the *Cork Examiner* called 'Carbery's Column'. He had been a hurler, and his brother Mick captained the Cork 1911 All-Ireland football team. Mehigan was the voice of the first live broadcast of a sports event and

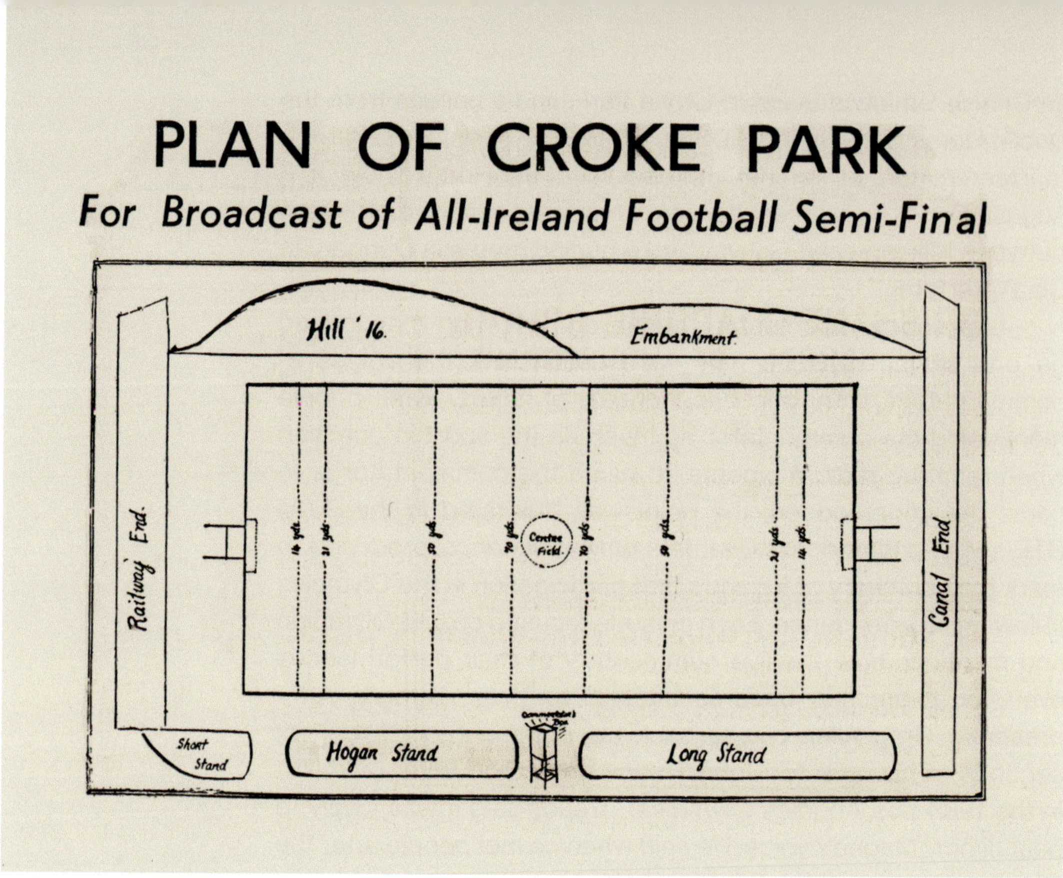

1935 Plan of Croke Park for listeners following the commentary on radio, published in the *Cork Examiner* of 17 August 1935.

the first by Irish radio. The commentary was of the All-Ireland Hurling Semi-Final between Kilkenny and Galway on 29 August 1926. The match time for the live broadcast was outside the usual programme times, so there were announcements made on air in the preceding days. Mehigan's radio commentary became very popular. As early as October 1926, the *Irish Independent* was reporting about the irreverent parodying by well-known comedian Jimmy O'Dea of the Mehigan style of commentary.

Éamonn de Barra, who would follow as a commentator on Radio Éireann, remembered Cork playing Tipperary in a Munster hurling championship final and that, due to widespread unemployment, many couldn't afford to travel to Thurles to see it. Brabant's, suppliers of wireless sets in Cork's Grand Parade, used to put a radio set with big gramophone-like horns out through the first floor of the office to broadcast the matches. Crowds would be there listening and looking up at what de Barra described as 'this bugle-like arrangement'. He remembered someone taking off his cap and throwing it up to the wireless and yelling out to a player what to do – people shouting at the radio seems to have been a thing since radio began.

The Tailteann Games took place in Dublin for the first time in 1924, though the idea to stage them had been talked about since before the Civil War. They originated from the idea of forging pride in ancient traditions of Irish heritage – like the recently revived Olympics, but with an Irish twist. World heavyweight boxing champion, Irish-American Gene Tunney had come to see the games in 1928. He was the first international sport celebrity to be interviewed for the radio station, something as exciting for some as listening to – or indeed now watching – a sports person in full flight. It was Mehigan who got hold of Tunney to say a few words to 'the folks', as Tunney called the listeners and those present in Croke Park.

Soccer's first commentary on 2RN was on the semi-final of the Free State Cup between Drumcondra and Fordsons. In a letter published in *Football Sports Weekly* on 26 February 1928 entitled 'One Who Listened-In' the writer complained about the announcer and their perceived lack of knowledge about the game: 'If some of our reporters were to describe a match in that manner in a newspaper, what would happen to the reporter? There was, however, much humour for me and my female friend in "listening in" to this game; but we were sadly disappointed in not getting a better and more actual description of the "doings".'

In 1930 *Gaelic Sports News* was introduced to the radio schedules on Sunday evenings. Sean Callaghan, who became known as Seán Ó Ceallacháin, set up a system for results from club matches to be telephoned to him from around the country. In 1953 Seán Óg Ó Ceallacháin took over the programme from his father until his own retirement in 2011. Ó Ceallacháin entered the 2012 *Guinness Book of Records* for achieving the longest career in radio sports broadcasting worldwide. His first radio broadcast took place on Radio Éireann in 1935 as part of Dublin's Scoil Colmchille primary school choir.

When RTÉ Radio was fifty years old in 1976, Roibeárd Ó Faracháin, by then retired controller of programmes, pointed

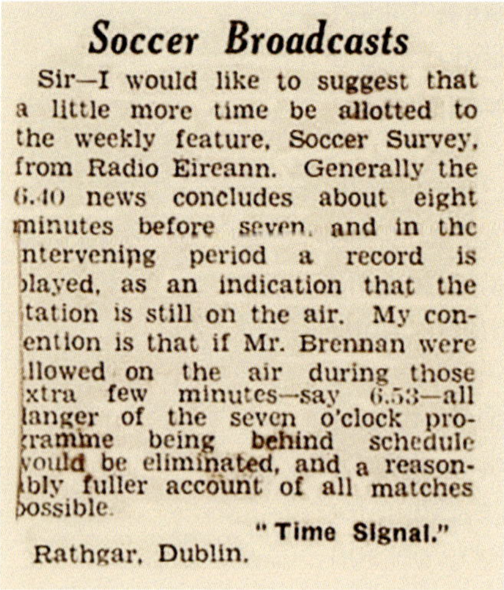

Soccer Broadcasts

Sir—I would like to suggest that a little more time be allotted to the weekly feature, Soccer Survey, from Radio Eireann. Generally the 6.40 news concludes about eight minutes before seven, and in the intervening period a record is played, as an indication that the station is still on the air. My contention is that if Mr. Brennan were allowed on the air during those extra few minutes—say 6.53—all danger of the seven o'clock programme being behind schedule would be eliminated, and a reasonably fuller account of all matches possible.

"Time Signal."

Rathgar, Dublin.

Letter in the *Evening Herald* of 21 September 1946 requesting that more time be given to soccer broadcasts on Radio Éireann.

out the enormity of the operation of collecting results that Ó Ceallacháin had single-handedly established for listeners. Tuning in for results on Sunday evenings became a weekly ritual long before they were instantly checkable online.

Gaelic Sports News only covered GAA, so listeners who followed other sports were understandably none too pleased with the concentration on one organisation's sports on the one programme on the station. A letter published in the *Irish Independent* on 22 August 1933 went on to say: 'I beg to protest against the action of the broadcasting authorities in omitting from the sporting news on Sunday any reference to the results of the games played under the Association Football code. Now that the soccer season is opening, I hope they will give followers of that code an equal opportunity to get the results of the game in which they are interested.' For a while that happened, but it soon reverted to GAA-only coverage.

Not everyone was happy with the standards of GAA commentary. An enthusiast in Ardara asked in the *Irish Independent* of 14 August 1933:

> Could some improvement be made on the running commentary on Gaelic Matches. Many of us here in the North are very interested in Gaelic matches and can only hope with the aid of a good commentary to visualize the play. Can't we have announced to us the flight of the ball in its principal movements, movements that count in the game? Can't we be told who plays the ball, to what side frees etc. are given, who takes them and the result?

An incident during a match illustrated the country's still-volatile state in the aftermath of independence and the civil war and showed how widely sports broadcasting was understood to reach audiences. The Fianna Fáil government had been in power for only a little over a year. Éamonn de Barra was broadcasting that year's 1933 All-Ireland Football Final from Croke Park when an armed intruder took over the microphone and made a demand for republican prisoners in nearby Mountjoy Gaol to be released. It was all over in about two minutes when de Barra picked up again to complete the match commentary. The *Irish Independent* of 25 September reported how thousands of radio listeners experienced what was described as 'an unexpected thrill'.

In the 1930s the station broadcast live relays from the FA Cup Final in Wembley and the Grand National in Aintree. Billy Hooker gave soccer's first live commentary from Dalymount in 1935, and Mick Byrne broadcast the first commentary of an Irish motor race, the Derby of 1936. The microphone picked up the racing cars in the Phoenix Park. Illustrating both the scarce resources and inventive make-do of those involved in early sports broadcasting to offer so much variety to listeners, an old post office truck was used to catch the race, with the commentator in the cab and the technical equipment in the back – and during bad weather too. Four decades later, in 1979, RTÉ Radio marked fifty years of the Irish Grand Prix of motor racing in the Phoenix Park when president of the executive council of the Free State, W. T. Cosgrave, started the race. The prize was £1,000 and a twenty-inch-high replica of the Wellington monument in solid silver on a Connemara marble base.

Relations between Radio Athlone (as it was called at the time after the location of the new transmitter) and sporting organisations were sometimes volatile. There was no broadcast on St Patrick's Day 1936 of the Annual Railway Cup because of differences of opinion on the commentary between the station and the GAA. That year's All-Ireland Football Semi-Final was in Killarney, with no resolution reached between them about coverage of the match. Radio Athlone made its own arrangements to bring the match to listeners. Seán Ó Ceallacháin watched the first half and ran to a telephone to broadcast what could be described as a delayed live commentary. Éamonn de Barra did the same for the second half.

Cumann Lúiṫ-Ċleas Gaeḋeal.

Craoḃ na h-Éireann sa Peil

i bPáirc an Ċrócaiġ
Meaḋon Foġṁair a 24, 1933

Caḃán
v.
Gaillimh

ar a 3.15 a ċlog

Réiḋtiġteóir—
Máirtín Ó Néill

Clár Oifigeaṁail

Leaṫ-Ċraoḃ na h-Éireann, Peil na nÓgánaċ

Aontruim
v.
Ciarraiḋe
2 p.m.

Réiḋtiġteóir—Seán Ó Duḃgail

Pádraig Ó Caoiṁ

Dul isteaċ: 1/- agus 2/-
Breis le díol ar Ṡuiḋeaċáin ar an taoḃ-líne agus ar Árdán Uí Ógáin.

Luaċ - - - - - 2p.

Wood Printing Works Ltd., Stafford St., Dublin.

The programme of the 1933 All-Ireland Senior Football Final between Cavan and Galway. The live radio commentary by Éamonn de Barra was briefly interrupted by an intruder who took the microphone demanding that certain prisoners be released, before de Barra went on to complete the commentary.

The Irish Press of 25 September 1937 reported that, for the first time in the history of Irish sport, a priest would be the radio commentator in a football match. He was the Reverend Father Hamilton of Ennis, Co. Clare, who would broadcast the All-Ireland Football Final due to take place the following day at Croke Park between Cavan and Kerry. This was the same Canon Hamilton whose suggestion that a major match in 1947 between Cavan and Kerry be played in New York – and the match did indeed take place – more on that later.

The year 1938 was significant in GAA and Radio Éireann history. Michael O'Hehir, at seventeen, applied to be a match commentator. His live audition – along with that of several others – was in Croke Park during the broadcast of the National League match between Wexford and Louth. O'Hehir's gift was immediately recognised and Dr T. J. Kiernan, then Radio Éireann's director, instructed him to commentate on the entire second half by himself. His first full live match broadcast was in Mullingar in that year's Football Championship Semi-Final in August, when Galway beat Monaghan.

Speaking in 2001, Frank Corr, former Head of RTÉ Sport, said:

> Michael O'Hehir gave the impression that he was talking to you all the time and he had all these phrases that he used. He personalised the game. Up to that point the commentator gave you the facts of what was happening, but he didn't perform. What O'Hehir did was to bring performance into it. He always tried to keep the listener listening by persuading him that the match was either better than it was and the team they were supporting had a better chance than they had.

The Second World War years saw a huge surge in listenership for live sports broadcasts of field games, indoor sports and water sports. As the war progressed, sports and matches continued to be played, though travelling became more difficult when fuel was rationed and trains were running on wood. Michael O'Hehir began to do racing commentary in 1945. The 1947 Galway Plate is the earliest recording that the RTÉ Archives has of his racing commentary. The war years presented multiple technical challenges. The restrictive finances of the station meant purchasing necessary equipment was prohibitive. It also meant

Clockwise from top: ticket to the 1947 All -Ireland Gaelic Football final at the Polo Grounds in New York; the Kerry team en route to New York; the victorious Cavan team in their dressing room; a promotional postcard for the event.

The Cavan homecoming after they beat Kerry 2–11 to 2–7 in the Polo Grounds, New York.

resorting to renting microphones on a costly weekly basis from Standard Telephones. Though Radio Éireann was a regular customer, the company insisted on a flat charge of ten shillings for each microphone. Supplies for repairs was equally difficult, with parts scarce.

The year 1947 was also the centenary of Black 47 – what is often considered to have been the worst year of Ireland's Famine and the cause of mass emigration to North America. To mark the occasion, it was decided to play that year's All-Ireland Football Final, between Cavan and Kerry, in the Polo Grounds in New York; Canon Hamilton's suggestion of doing something in America had caught the public imagination and finally come to pass. Cavan emerged the victors with a final score of 2–11 to 2–7.

Passion and appreciation for the GAA varied considerably among decision-makers in the department of posts and telegraphs and consequently their willingness to facilitate what was required for the groundbreaking New York broadcast. While

Séamus Brennan, minister for posts and telegraphs, was an accomplished footballer, Seán Moynihan of the department of finance asked if anyone was interested in listening to football matches. He was proved wrong; that evening (because of the time difference) every car in O'Connell Street in Dublin was parked as people crowded around radios to listen to the game. Michael O'Hehir provided his live commentary to 'the folks back home', as he called them on air. The match had all the drama of the best big games, including a technical one: it overran and, but for the fact that New York's Mayor Bill O'Dwyer was from Bahola, Co. Mayo, extra time on the wires booked for the broadcast might not have been possible. The diaspora went into action in New York while bonfires were prepared in Cavan to welcome the team home with the Sam Maguire.

In 1949, at the age of eighteen, Micheál Ó Muircheartaigh started his broadcasting career, providing the commentary in Irish for that year's St Patrick's Day Railway Cup football final between Leinster and Munster. Radio Éireann's *Sports Stadium* programme began in that year too. It would, in time, move to television – often the acknowledgement of the success of a radio programme.

Following on from the success of the New York broadcast in 1947, the GAA and Radio Éireann reached an agreement to do the same for the 1951 National League Final. The department of posts and telegraphs, however, were not going to spend money on sending O'Hehir to New York again, and instead insisted that Radio Éireann hire an American commentator (of Irish descent), John 'Lefty' Devine. The department perhaps did the station an accidental favour, by showing up, as if it were needed, the star quality of broadcaster Michael O'Hehir. The *Radio Review* of 4 October 1951 published this about Devine's commentary: 'The first job of a commentator is to tell his listeners what is going on. His second to relay and sometimes to create the excitement of what is going on and his third – to keep talking at all costs. Mr Devine did none of these things. Instead, he gave us a raft of irrelevant details. Interpolated into a commentary that will be memorable for the endless chain of pauses. And his full wanton wanderings from the point of play.'

Sports viewing options from the holiday features on Irish radio and television as listed in the *RTÉ Guide* in 1962 on the first St Patrick's Day since the introduction of Telefís Éireann.

HOLIDAY FEATURES:

Radio Eireann

Mar aduirt Padraig fein:

Caint o'n **Athair Eoghan O Frighil,** bunaithe ar an bhFaoistin agus ar an Litir chun Choroticus.
12 noon

Bliain Phadraig:

Events of the Patrician Year recalled by **Sean Mac Reamoinn.**
12.45

The Irish on parade

New York goes gay on Ireland's national holiday. **Stephen Rynne** talks about how the Irish—and others—salute our Patron.
7.15

Festival Choice:

Ciarán Mac Mathuna presents traditional music which listeners —many abroad — have requested.
7.30

Coirm Cheoil na Feile:

Ceoil Fhuireann Siamsach Radio Eireann: Cór Radio Eireann: Mary O'Hara, William J. Power, Micheál O Conaill, Darach O Cathain.

Ceoltóiri Chualann á stiuradh ag **Seán O Riada.**

Rinceóiri Cora Cadwell le Gráinne Nic Chormaic. Stiurthoir: **Eimear O Broin** Compere: **Niall Toibin.**
9.0

SPORT 2.25—5.0

Railway Cup Finals: At Croke Park.

Iomanaoicht: Laighean v. Mumhan. Cúntas reatha le Micheál O Muircheartaigh.

Football: Leinster v. Ulster. Commentary by Liam Campbell.
(Athlone Wavelength: 530 m.)

Soccer: League of Ireland v. Italian League. At Dalymount Park. Commentary by Philip Greene; summaries by Kieran Kenealy.
(Dublin and Cork Wavelengths)

Telefis Eireann

SPORT

Our Outside Broadcast cameras visit Croke Park for the Railway Cup Finals: Viewers will see the second half of the hurling final, **Leinster v. Munster,** and the whole of the football final, **Leinster v. Ulster.** Commentaries by **Micheal O Hehir.** Produced by **Philip Thompson.**
3.15

Goltrai agus Geantrai:

Seamus Ennis introduces a special programme of Irish music, song and dance. With **Annraoi O Raghallaigh** and **Padraig O Broin,** dancers; **Breandan O Dubhghaill** and **Aideen Collins,** ballad singers, and fiddler. **Sean Bracken.**
5.20

Glamour of Galway:

A film of the Gateway to the West and its surroundings
6.45

An Roisin Dubh

The legend, laughter and sadness of Ireland present and past are woven into an evocative pattern of dance, music and song. Special choreography directed by **Norman Maen,** with settings by **Alpho O'Reilly**
8.30

St. Patrick's Night: Television Ceili:

Sean O Siochain is Fear a' Tighe at a ceili brought to you in an Outside Broadcast from St. Francis Xavier Hall, Dublin.

Music by the **Dermot O'Brien Ceili Band.**

Figure dancing by the **Harry McCaffrey Senior Dance Group.**

With **Seán O Sé;** harpists **Mairin Ni She, Peigi Ni Maoileannaighe, Aoife Ni Chiosáin;** pipers, **Leo Rowsome, Sean Seery, Jim Dowling** and **Eugene O'Reilly.**

Produced by Philip Thompson (See Page 8).
9.15

As the department of posts and telegraphs gave a greater measure of independence to the station during the 1950s, Radio Éireann increasingly considered how listeners received broadcasts, focusing on both content and presentation. These included navigating the inclusion of the performance of 'God Save the Queen' before international rugby matches (playing the British national anthem did not sit comfortably with many listeners or staff yet); the sports to be prioritised for broadcast from Athlone; using the wider-reaching, better-quality transmitter option; and how to include the Irish language in sports broadcasting.

A dramatic moment in Irish soccer commentary came in 1957, when Philip Greene described a last-minute England goal at Dalymount that knocked Ireland out of World Cup contention, leaving listeners crestfallen. It has echoes for later generations of Scalisi's goal in 1990, Thierry Henri's 'Hand of God' moment in 2009, or when, more recently, Ireland's hope of qualifying for the 2026 World Cup slipped away.

By 1958 Radio Éireann had started its end-of-year annual *Sports Review*. Irish golfers Harry Bradshaw and Christy O'Connor were key contributors to the programme, having won

Broadcaster Jimmy Magee in 1966, who by then was known as the 'Memory Man' for all his sporting facts.

golf's Canada Cup for Ireland in Mexico City. The tournament would be held in Portmarnock in Dublin in 1963, when Radio Éireann covered the event.

In 1960 Radio Éireann travelled, for the first time, to bring Ireland its own coverage of the Olympics in Rome. This had not been the case when Ronnie Delany won Ireland's first Olympic gold medal for athletics in Melbourne in 1956. There is, however, a Radio Éireann interview with Delany, then aged twenty-five, broadcast at the time of his preparations for the Rome Olympics – he was in conversation with, curiously, the Irish tenor John Feeney. The interview begins with the two men talking about stage fright and the need, Delany says, to have a certain stage fright to perform. To Feeney's question about what he eats before a race, Delany said he starts off with a fruit cocktail followed by a big bowl of chicken soup. After that a small steak, a couple of helpings of vegetables and another fruit cocktail before finishing with a cup of tea. He talks about having taken a Master's in Theatre Arts in the past year and that he had done some acting and a little producing. Asked about his hobbies he mentions hi-fi equipment and a fondness for classical music and Irish folk songs. He explains that he is currently working with Irish International Airlines in the US as a salesman. Although the conversation sounds a bit staged – perhaps with an international audience in mind for St Patrick's Day – it must have been a thrill for Irish listeners to hear Delany speaking on the radio in such a way at the height of his fame.

Noel Andrews (in headset) commentating for RTÉ Radio at the European Boxing Championships, Bucharest, Romania in 1969.

Reporting of world sport continued to evolve on the station. The *RTÉ Guide* of 21 September 1962 reported that Radio Éireann would make history when, on Wednesday morning at 3.45 a.m., Noel Andrews would broadcast live from the world heavyweight championship fight in Chicago between champion Floyd Patterson and challenger Sonny Liston. On 22 October it reported that, at 6.55 a.m. on Wednesday morning on Radio Éireann (Athlone wavelength), Phillip Greene would give the commentary from Seville on the World Cup-qualifying match between Ireland and Spain.

What happened in 1963 must have made up somewhat for the last-minute defeat of Ireland by England in 1957. The League of Ireland, in what was described as their most glorious night, beat the full England team 2–1: as Philip Greene remarked, 'look at that invasion of Dalymount'. That year too, the Sam Maguire was brought over the border for the first time when Down beat Kerry 2–10 to win the All Ireland Senior Football Championship Final.

Philip Greene, Dermot Kelly and John Comyn provided commentary from the 1972 Munich Olympics in what was then West Germany.

(Left)
In 1976 a cross-generational cohort of Irish competitors vied for the £35,000 prize in the Irish Open. The *RTÉ Guide* had profiled the relatively unknown American winner Ben Crenshaw ahead of the event.

(Right)
The *RTÉ Guide* cover for the 1972 All-Ireland Hurling Final featured Alan Corsini's picture of the 1971 finalists!

The arrival of Raidió na Gaeltachta in 1972 brought with it a further independence, a *cluas eile* (another ear), to paraphrase TG4's *Súil Eile* (Another Eye), to RTÉ radio broadcasting. As Irish radio approached its fiftieth year, it increasingly recognised that sport, including the GAA, should not be reported on or commentated solely by men. Trailblazing sports broadcaster Doirín Nic Mhurchú travelled on her Honda 50 to report from the early years of RnaG. 'This is for the press, missus,' Nic Mhurchú was told whenever she arrived at a match. 'I *am* the press,' was her answer. Nic Mhurchú became an authority on all sorts of sports when she began working on *Spórt an Domhnaigh*.

The launch of Radio 2 introduced other voices relating to sport as coverage of it extended across two stations now. There was Caroline Murphy, as well as Michael Lyster, who joined the station as sports-news anchors. In their first summer on air, Radio 2 DJs Vincent Hanley and Ronan Collins broadcast from the Galway races in a mobile studio by the racetrack.

Mick McCarthy and members of the Republic of Ireland squad on O'Connell Street for a homecoming reception on 1 July 1990, after getting to the quarter finals of the FIFA World Cup Finals in Italy.

Italia 90 was major. It was the first time the Republic of Ireland qualified for the World Cup. RTÉ Radio concentrated on the Irish team matches. *Sports around Seven* on Radio 1 came from Cagliari and Palermo on three consecutive Fridays presented by Des Cahill and Gabriel Egan. Martan Ó Ciardha reported for RnaG, Joe Duffy for *The Gay Byrne Show* while Barbara Jordan and Brenda Donohue were in Italy on behalf of *The Gerry Ryan Show* on 2FM. Alex Ferguson, then Scotland manager, offered prophetic words in the *RTÉ Guide* of 8 June 1990:

> Ireland's strength undoubtedly lies in midfield. Players such as Paul McGrath, Ronnie Whelan, Ray Houghton and Kevin Sheedy come tripping off the tongue and there are a number of very, very good midfielders in the panel, among them Andy Townsend. Another plus in favour of the Irish is that they don't give anything away at the back, which is why I think the side is going to do very well in the finals ... they might just be the dark horse of the Italian World Cup.

▶
World champion featherweight boxer Barry McGuigan was set to defend his title on 14 February 1986. Jimmy Magee, who had covered all his fights until then, was ringside for Radio 1. McGuigan beat Cabrera in the RDS but would lose to Steve Cruz in Las Vegas in June 1986.

(Opposite page)
The first RTÉ Radio 2 Dublin City Marathon started on St Stephen's Green South on 27 October 1980.

(Opposite page)
Winners of the Radio 2 Dublin City Marathon, on St Stephen's Green on 28 October 1985. Left to right: Julia Gates (first woman), Gerry O'Rourke (first wheelchair competitor) and Dick Hooper (first man).

Gold Medallist Kellie Harrington with the Ireland boxing team at Dublin Airport on return from the Tokyo 2020 Olympics.

RADIO 2 DUBLIN CITY MARATHON
Smurfit Computing
SEIKO
QUARTZ TIMING
Hertz
Dublin Corporation
COSPOIR
CAPE SCAFFOLDING
SEIKO TIMING
RTE RADIO 2
DUBLIN CITY MARATHON

A view of participants performing a choreographic routine at the opening ceremony of the Special Olympics, in Dublin's Croke Park on 21 June 2003, the first time the Games were held outside the US.

The Special Olympics opened in Dublin's Croke Park on 21 June 2003. It was the first time the games – in which athletes with intellectual disabilities compete against one another – was held outside of America, with 7,000 athletes from 160 countries taking part over the course of ten days. RTÉ Radio, along with RTÉ Television, broadcast live coverage of the opening ceremony.

Ireland's first woman to host a radio sports programme was Jacqui Hurley, who became the anchor of *Sunday Sport* on Radio 1 in 2009. Hurley, with Marie Crowe, would go on to co-present *Inside Sport*, Radio 1's weekday sports programme, launched in 2025. In 2023 Eimear Ryan spoke on Radio 1's *Ray D'Arcy Show* about her new book, *The Grass Ceiling*, drawing attention to women in sport from the inside out. As well as being a writer, she also played camogie for Co. Tipperary. She recalled how, in 2004, the centenary of the formation of the Camogie Association, there was talk of the final being the curtain-raiser for that year's All-Ireland Senior Hurling Final, which at the time would have brought visibility to camogie. It was thought that the minor teams could not be displaced for the women – even though it was their centenary. She talked about the huge improvement in media coverage in women's sports – particularly

Left to right: Commentator Pauric Lodge, *Sunday Sport*'s Jacqui Hurley and Con Murphy for the 2012 Hurling All-Ireland Senior Championship Final Replay, Kilkenny v Galway in Croke Park.

Micheál Ó Muircheartaigh, setting the scene for his commentary of the 1987's Munster Senior Hurling Final featuring Cork v Tipperary in Semple Stadium, Co. Tipperary.

women's games being aired live – which allowed people to buy into the game and to get to know the players, the teams, the dramas and the rivalries: to access the narrative the way men had been doing with men's sports for decades.

In recent decades, evolving technologies have enabled global audiences to more easily access Irish sport while also allowing them to directly engage with RTÉ. Comments from listeners have included, for example, that it was 'unnecessary' and 'spoiled Michael Ó Muircheartaigh's informative commentary' to have a second person commentating with him; that it was infuriating when the sports commentary switched from one

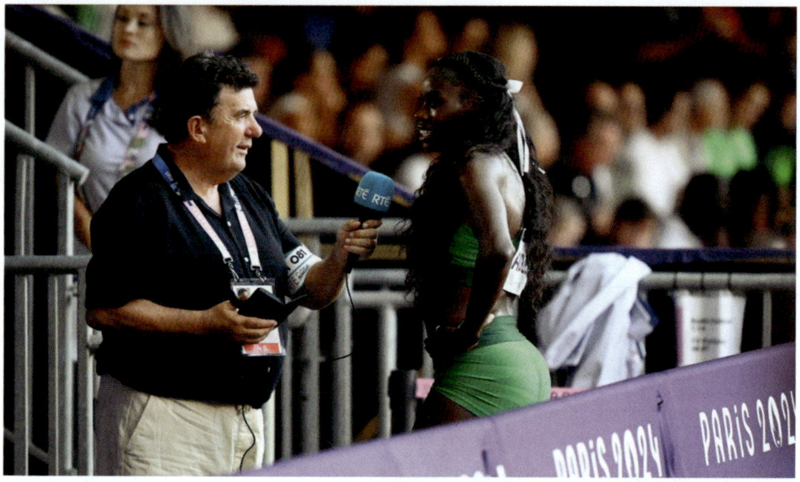

Rhasidat Adeleke is interviewed for radio by RTÉ's Des Cahill, after the women's 400m final at the Stade de France on 9 August at the 2024 Paris Olympic Games.

Team Ireland's rowers return from the Tokyo 2020 Olympics, Gold medallists Paul O'Donovan, left, and Fintan McCarthy alongside bronze medallists, from left, Emily Hegarty, Fiona Murtagh, Aifric Keogh and Eimear Lambe.

major fixture to catch up on progress on another; that it was a privilege to broadcast GAA games; that RTÉ should advertise its programmes more in the UK; what is the correct way to pronounce the name José; what is the correct plural term for euro; or simply a request for a mention for himself and his wife on *Sunday Sport* – after all, there's a thrill in hearing your name on the radio.

A chapter on sport people heard on RTÉ Radio during its first century could only ever nod to a fraction of that history. The chapter is also, inevitably, somewhat lopsided in its representation of women, because that reflects the reality for much of the period covered. Wasn't the RTÉ sports broadcaster Philip Greene at least truthful when, towards the end of his 1976 radio programme, *This Game is Not Over Yet: Fifty Years of Radio Sport*, he said: 'These have been some of the moments, some of the men, some of the memories which have filled our first fifty years'? Indeed, the game isn't over. Radio continues to evolve, bringing us sport of genius and disappointment. Hopefully, what has been included here will also bring to mind many of the other radio sports moments from the same period.

Ireland rowers, from left, Aifric Keogh, Eimear Lambe, Fiona Murtagh and Emily Hegarty celebrate after finishing 3rd place in the Women's Four final at the 2020 Tokyo Olympics

SEAMUS HEANEY

Heaney photographed in the *RTÉ Guide* in 1972 introducing *Imprint*.

No one encapsulates the attraction and potential of radio in quite the same way as Seamus Heaney does. Heaney had a sure fondness for radio. He said as much in his Nobel prize acceptance speech in 1995, when he recalled hearing the radio 'alive and signalling' in his childhood, and how, in time, he grew to listen more deliberately, perched on the arm of the sofa at home so he could hear it properly.

From the late 1960s until his death in 2013, Heaney regularly contributed to Radio 1. 'I do enjoy being a disc jockey, even for a quarter of an hour,' was how he once merrily signed off his weekly programme *Poems Plain* in 1978.

In 1972 Heaney presented the book programme *Imprint*. He had recently moved to Co. Wicklow to concentrate on his writing. In interview between Heaney and Dennis O'Driscoll in 2008 he said that being heard on the radio by the customers in his local pub helped 'settle us in'. Like most of us, having an income was important: 'I was anxious to have some kind of regular money coming in, so the weekly fee was reassuring.' In the *RTÉ Guide* he described the earlier poetry programme *Anthology*, presented by Austin Clarke, as a key influence on him and on *Imprint*. It was a typically affirmative nod by Heaney to the fact that most radio programmes have evolved in some way from antecedents: 'here was a man [Austin Clarke] discussing and talking about something he cared about and wanted to share. *Imprint* sets out to be a programme like that: a forum where people who know and care about books talk about reading and writing them ... it's a programme for those who actively want to listen'.

For the next thirty years Heaney clocked up several further hours on the radio – interviews, conversations, poetry, scripts and radio talks – and continues to be the most quoted poet on RTÉ Radio of the last hundred years. On 6 October 1995, the day after Heaney's Nobel Prize announcement, David Hanly on *Morning Ireland* mischievously joked: 'They seek him here, they seek him there but so far no one has managed to find Ireland's fourth Nobel Laureate for literature, Seamus Heaney, in fact since he is still holidaying in Greece it is still possible that he doesn't even know that he's won!' He managed to track him down for the programme the following day.

Seamus Heaney in RTÉ Radio in June 1987, when he was Colm Tóibín's guest on *The Poet's Voice*. His latest poetry collection, *The Haw Lantern*, had just been published.

▼

Seamus Heaney with his family: (from left) his son Michael, daughter Catherine Ann, wife Marie and son Chris, after receiving the Nobel Prize for Literature in Stockholm on 10 December 1995.

Radio 1 broadcast a celebration of his seventieth birthday on 13 April 2009. It was a day-long on-air affair, mostly from the Irish Museum of Modern Art. John Kelly welcomed radio and online listeners around the world: 'We're here in many ways to thank him, not just for the work which has nourished us, and delighted us, and moved us, and steadied us throughout the years, but also for the way he has carried himself and carried us.'

(Left)
Seamus Heaney following the birthday speech he gave at the 'Heaney at 70' event.

(Right)
Broadcaster and poet John Kelly embraces Heaney after his birthday speech.

Heaney's death on 30 August 2013 was mourned worldwide. The day's broadcasting on RTÉ turned to him first through *Liveline*, presented that day by Philip Boucher Hayes. The tributes poured in. RTÉ Radio had so much to be grateful for in his generosity; the many times we mithered him with yet another request to contribute to a programme one of us was making. He so often said yes.

His funeral Mass on 2 September was broadcast on RTÉ Radio and coverage streamed worldwide. Radio 1 Extra carried a continuous broadcast of the poet reading his poems from the *Collected Poems* CD, produced to mark his seventieth birthday, but which in truth he had gifted the future.

The day before the funeral I had been in Croke Park, for the Dublin v Kerry Football Senior semi-final. On these big match days GAA members whose deaths have recently occurred are honoured with a few brief words, followed by a minute's silence. When Seamus Heaney's name was announced the packed stadium stood and applauded before the announcer's voice picked up again: 'Seamus played underage football with Castledawson, and his work shone a leading light on his rural area of Bellaghy and Co. Derry and on the life of our nation as a whole. *Suimhneas síoraí ar a n-anamnacha uaisle.*'

There it was. The capacity for the world to become local in an instant; a community in seconds, acted out in the stadium. As I watched the teams already out on the pitch below I thought how, like so many present and tuning in to listen on all sorts of devices, in Ireland and across the world, they too would have been taught his poetry at school, or heard teachers, or family or friends, or public figures turn to his words to express what they couldn't. Most of those gathered to play, watch or listen that day, if not familiar with the lines of his poem 'Markings', certainly knew their emotion:

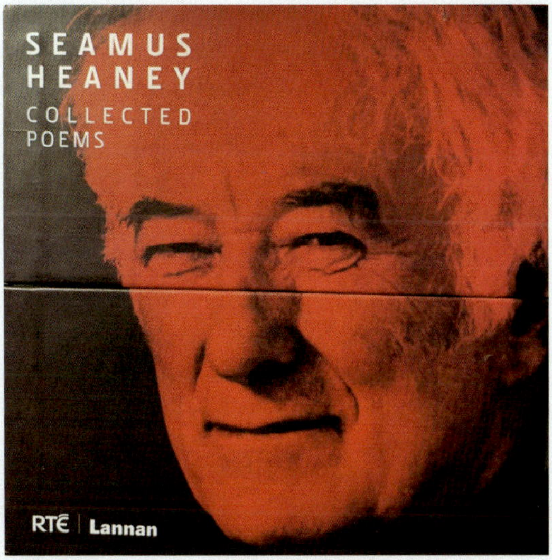

The 2009 CD box set *Seamus Heaney Collected Poems,* produced by Tim Lehane with an introductory text by Peter Sirr.

> Youngsters shouting their heads off in a field
> As the light died and they kept on playing
> Because by then they were playing in their heads.

Croke Park, 1 September 2013, where the packed stadium burst into applause on the poet's name being mentioned before the match started, captured here on a mobile phone.

ACTING OUT

CHAPTER 3

The Dublin station 2RN opened on the first day of the tenth anniversary year of the Easter Rising, which brought its own drama. The premier of Sean O'Casey's *The Plough and The Stars* at the Abbey Theatre (which had, the previous year, received its first state annual grant) provoked riotous reaction due to his representation of the despair 1916 had brought the city's citizens. There was the first fictionalised on-screen telling of the War of Independence – audiences loved Irish Destiny's romanticised account of the war. Conradh na Gaeilge/ Gaelic League branches had theatre groups and amateur drama groups in English also started taking root. All these fed an interest by professional and amateur performers and writers to get involved in drama on the radio, even though the single studio above a welfare office and the minuscule budgets for plays continued for some years.

In 1928 2RN moved from a temporary studio at 36 Little Denmark Street into the General Post Office. It was from the GPO studio that radio drama would find its voice and become a key contributor to the popular soundscape of mid-twentieth-century Ireland.

There were only three studios to begin with (though the area the station occupied in the GPO expanded over the years). The largest studio was where plays – and occasionally even opera along with varieties of music and songs that made up most of the early daily schedules – were broadcast live. Members of the modest and fancifully named 'station orchestra' would sometimes accompany plays, underscoring spoken descriptions of settings, characters, and so forth. This was before it became clear that effective radio drama works best when information is embedded within the script, or when literal settings are abandoned and left to the listener's imagination.

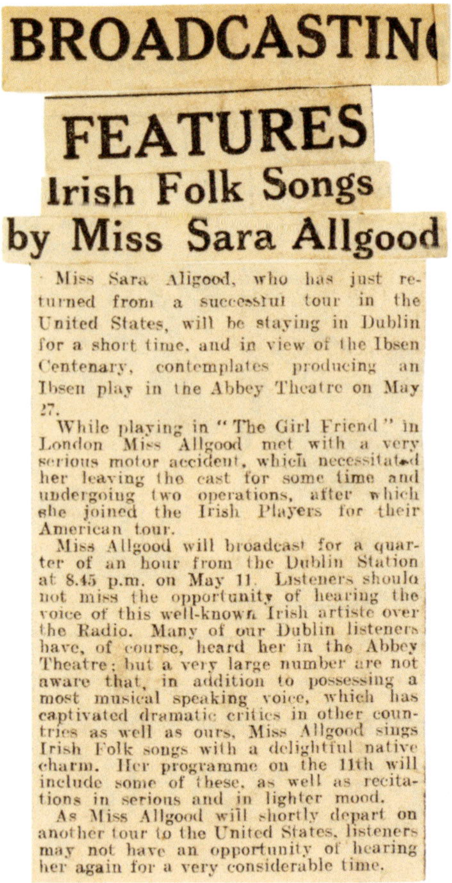

BROADCASTIN(

FEATURES

Irish Folk Songs

by Miss Sara Allgood

Miss Sara Allgood, who has just returned from a successful tour in the United States, will be staying in Dublin for a short time, and in view of the Ibsen Centenary, contemplates producing an Ibsen play in the Abbey Theatre on May 27.

While playing in "The Girl Friend" in London Miss Allgood met with a very serious motor accident, which necessitated her leaving the cast for some time and undergoing two operations, after which she joined the Irish Players for their American tour.

Miss Allgood will broadcast for a quarter of an hour from the Dublin Station at 8.45 p.m. on May 11. Listeners should not miss the opportunity of hearing the voice of this well-known Irish artiste over the Radio. Many of our Dublin listeners have, of course, heard her in the Abbey Theatre; but a very large number are not aware that, in addition to possessing a most musical speaking voice, which has captivated dramatic critics in other countries as well as ours, Miss Allgood sings Irish Folk songs with a delightful native charm. Her programme on the 11th will include some of these, as well as recitations in serious and in lighter mood.

As Miss Allgood will shortly depart on another tour to the United States, listeners may not have an opportunity of hearing her again for a very considerable time.

(Left)
***The Evening Herald* of 4 May 1928 reports that Sara Allgood, one of the most celebrated actors of her era, during a visit home will broadcast on 2RN.**

(Right)
Actress Sara Allgood.

Listeners may well have already heard plays on stations such as BBC or 2BE (later to become BBC Ulster and on air from Belfast since 1924) and some continental stations too. Those who listened to the wireless were well attuned to moving the dial purposefully or accidently to other stations on the same wavelengths. But here now were the country's own productions of plays on radio. These were linear presentations of stage plays, much like early television drama, which also borrowed from theatre. But what was significant was that the voices and sounds, scratchy as they may have been to begin with, were broadcast live for listeners to enjoy at home. The radio brought drama to many who had never been to the theatre. This intimate listening experience must surely be one of the reasons why, from the outset, there was such strong opinion on what was heard – never mind listeners sometimes believing that they personally knew characters familiar to them from the radio.

The station's next move was at the turn of the 1970s, when it relocated to the newly built Radio Centre out in the leafy suburban campus just beyond Donnybrook village – and where, almost a decade earlier, RTÉ Television had pitched its tent. Here radio drama would have a dedicated studio, an accompanying green room and the opportunity to make itself new again. The move aligned with a period during which radio generally was reinventing itself as a livelier, more relevant voice, finding its own distinctive strengths to challenge the hold television had established over its audiences.

So how did drama get to be on the radio? What usually happened was that a producer – or someone who called themselves a producer – came along with a script to the first station director, Séamus Clandillon, and got a booking for a studio in the live schedule. They would be paid a meagre fee to deliver their play, out of which they were required to cover every aspect of the production. In such straitened circumstances, producers often got people or groups they knew to perform. Rehearsals were minimal and sometimes non-existent. Newspapers promoted and reported on play and opera broadcasts. They sometimes included a photograph of casts, sometimes an amateur drama group of the Gaelic League or a school production in Irish or in English. From the first month of the station going on air, actor Frank Fay had given what became regular poetry recitals, which he gradually extended into presenting passages from classical plays. Alongside broadcasts from the station studio, listeners also began to hear actors in broadcasts of live relays from theatres such as the Abbey and the Gate – by actors including F. J. McCormack, Eileen Crowe and Sara Allgood – who took part in plays by writers such as Lennox Robinson, Teresa Deevy and Brinsley MacNamara, as well as translated international works. Abbey Theatre casts sometimes broadcast plays from the station studio too; the helpful continuity announcer was on hand to suggest, for example, that a particular play be listened to in silence and in the dark.

Because plays were not recorded until after the Second World War, actors could not hear themselves back and evaluate their work – the earliest existing recording of a Radio Éireann broadcast drama is *Interlude*, a short play by Paul Vincent Carroll from the late 1940s.

When the station's acting company was established in 1947, drama received a much-needed boost. The driver behind the Radio Éireann Repertory Company – later called the Radio Éireann Players and affectionately known as the RÉP – was

A Four-footed Actor.—"Mike," a Kerry Blue,' owned by Mr. W. McLysaght, of Kincora, Shandon Drive, Dublin, expects to be in fine barking form when he appears in the play, "A Change of Tempers," at 2RN to-night.

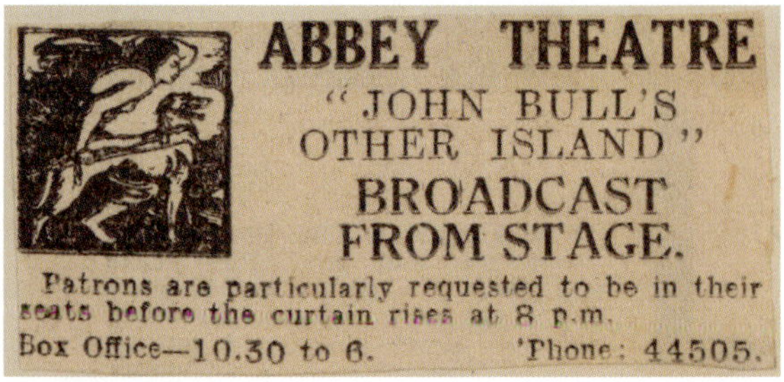

The Gorey (Wexford) Gaelic League Players who broadcast the Irish play, "An Feall," from 2RN last night. Giolla Criost O Broin, the author, is seated in centre of front row.
[I.P. Photo.

(Clockwise from left) 2RN delivered its own novelty casting in productions of plays, as this photograph and caption illustrate in the *Evening Mail* of 13 May 1931.

Live relays, or outside broadcasting of plays from the Abbey featured on 2RN.

Photograph from the *Irish Press* of 15 September 1935.

Roibeárd Ó Faracháin, a renowned poet and by then the station's controller of programmes. Micheál Ó hAodha, assigned as their first producer, along with Ó Faracháin and Ria Mooney, a producer in the Abbey Theatre, auditioned approximately 400 people for eighteen vacancies that fine-weather summer.

The recruited players, many of them native or fluent Irish speakers, had a peculiar status of employment: they were given the formal title of temporary, unestablished civil servants, because as actors they didn't fit into the employment structures of the department of posts and telegraphs under which the station functioned. Until 1973 women civil servants were required to resign from their employment in the civil service on marriage. However, Ó Faracháin insisted that women could remain after marriage because the RÉP required them for female roles. The number recruited for the RÉP, relative to the size of Radio Éireann, exceeded that of other similar public radio services. BBC had up to fifty radio actors at any time during the mid-century heyday of radio drama, but on short-term contracts.

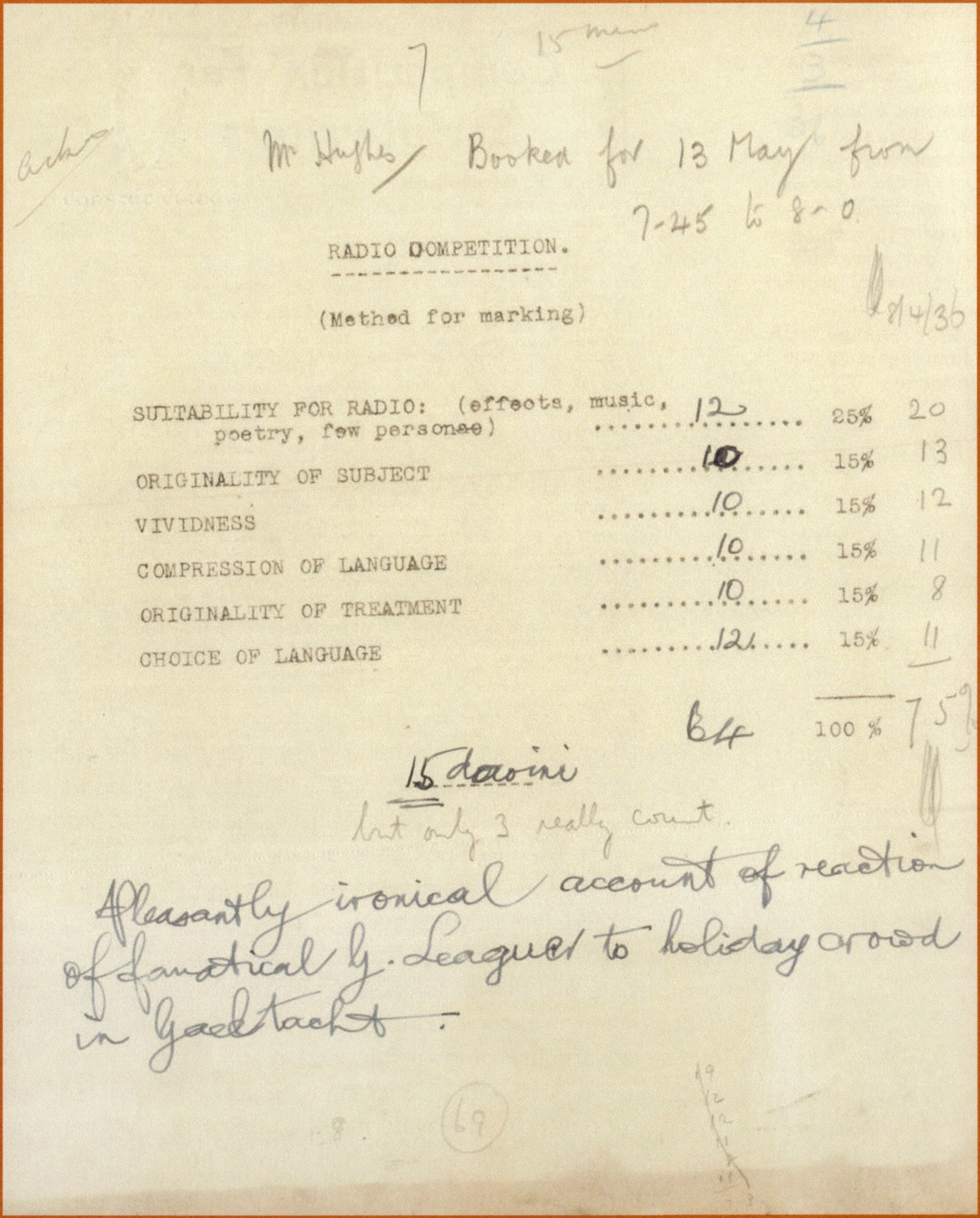

7 15 mins 4/3

Mr Hughes / Booked for 13 May from 7-45 to 8-0.

RADIO COMPETITION.

(Method for marking)

8/4/36

SUITABILITY FOR RADIO: (effects, music, poetry, few personae)	12	25%	20
ORIGINALITY OF SUBJECT	10	15%	13
VIVIDNESS	10	15%	12
COMPRESSION OF LANGUAGE	10	15%	11
ORIGINALITY OF TREATMENT	10	15%	8
CHOICE OF LANGUAGE	12½	15%	11
	64	100 %	75

15 daoini

but only 3 really count.

A pleasantly ironical account of reaction of fanatical G. Leaguer to holiday crowd in Gaeltacht.

A producer made notes on submission of a play script in accordance with the station's assessment criteria.

The search for radio scripts continued as illustrated in *The Irish Press* of 13 October 1947; and Brinsley McNamara's play Margaret Gillan and Julia Monks' adaptation of Jean Jacque Bernard's play *Invitation to a Voyage* involve frustrated, complicated marriages. Unlike printed literature, there was no censorship on radio plays.

Radio Commentary

Competition For Scriptwriters

By P. P. O'Raghallaigh

RADIO EIREANN'S new plan to overcome the script shortage is a new competition for short-story writers. Complete details will soon be announced on the radio, but I may mention that the prizes will be £15, £10 and £8 for the best three, while other suitable scripts will be accepted for broadcasting at the usual rates. Required scripts should be either 2,800 or 2,100 words in length (this is, of 15 or 20 minutes duration). In order to assist writers not accustomed to radio work, there will be a nightly talk from Radio Eireann in the week beginning October 26, in which valuable hints will be given.

These talks will be no less welcome to writers than the competition. Even writers accomplished in Press and magazine work show themselves unfitted for radio work through ignorance of its requirements. These talks should help to remove that disability.

An opportunity to compare R.E. and B.B.C. productions of the same play will be given to listeners to to-night's B.B.C. Home Service play, "Invitation to a Voyage." This is a longer adaptation of Bernard's play than the R.E. version, which was entitled "Mirage" and was played on Sunday and Thursday.

Julia Monks' R.E. adaptation was an exemplary script, well served by the production of H. L. Morrow and the four-voice cast—Aileen Harte, Ginette Waddell, John Stephenson and Ronald Ibbs.

Miss Harte played the part of Marie-Louise—a woman frustrated, married to one, dreaming of another, living in an atmosphere of business, craving for far-away hills—namely the Argentine and Philip. When the hills came to the prophetess, their greenness was proved a mirage and the dream vanished. The play was a study of the mind of Marie Louise and, in her interpretation of that part, Miss Harte deserves every credit. She made it an emotional experience, erring only in sometimes introducing emotion where it was not required. The other members of the cast played to perfection, and it is difficult to believe that to-night's B.B.C. production will impress more.

Saturday Feast

Last night's "Peter" will follow the new custom and be repeated on Thursday. Brinsley MacNamara's "Margaret Gillam" has been adapted for radio and is included in next Sunday's programme. It is another play which introduces matrimonial complications though, unlike "Mirage," the ending is tragic.

The coming week will not include many plays or features until Saturday. Philip Rooney's adventure series is continued to-night, and the Foley Family Cameos are due for another broadcast on Friday.

Saturday next will be quite a feast of light listening—County Clare is featured in the "Rambles Around Ireland" series; Green Room Nights and Microphone Parade take their usual place.

A new vaudeville programme is also fitted in—"Saturday Night at the Palace"—not an original title, perhaps, but a new venture for R.E.—favourite variety stars are included in the show.

From Belfast, the Irish Rhythms Orchestra and Ulste. Singers provide "The Music o Ireland" (the nation, not the composer) to-night. Programm is intended for listeners i France, announcements being i French. I hope the B.B.C. European Service will be the medium

To-morrow night's play fror Belfast, "Johnnie Comes Home, provides music of another natur —the roll of the Orange drun which are deeply welded into th plot.

Women made up many of the stronger new radio drama acting voices. Pegg Monahan was among the first to join, excited to be acting on radio – still a relatively new art form. She was eleven years old when she first listened to a radio play produced by Abbey actress Máire Ní Dhiúlaí. Ní Dhiúlaí required young boy actors for it and chose Pegg's brother, a pupil of hers at the time.

The ability to adapt to accents and characters was critical for the variety of demands made on RÉP actors. As well as learning how to deliver plays and poetry to a microphone, it was vital to have the capacity to move nimbly between Shakespeare or Synge, Laverty or Behan in any given week, not to mention being required to broadcast talks on miscellaneous subjects by people who were reluctant or considered not suitable to present their own words.

The by-now-familiar conventions of radio continuity to introduce a play could be exploited to best effect. P. J. O'Connor's 1966 adaptation of Bram Stoker's *Dracula* was introduced with the announcer saying:

> We advise listeners who are nervous or who may be alone or with children that parts of this programme are quite startling ...

The announcement was followed by a shudderingly long and harrowing screech, the sound of several knocks on a door creaking as it opens, heavy chains clanging, bolts opening and closing again, further creaking sound effects, silence ... and then a drawn-out voice slowly speaks:

> Welcome to my house, enter freely and of your own free will ... come freely, go safely and leave something of the happiness you bring ... I am Count Dracula, I bid you welcome to my house. Come in ...

In that production Peter Dix was Jonathan Harker and Thomas Studley was the count. *Dracula* was produced again in November 1997 to mark the centenary of the novel's publication, the 150th anniversary of Bram Stoker's birth and the fiftieth anniversary of the RÉP. This latter production was dramatised by Christopher Fitz-Simon and produced by Laurence Foster with actors Deirdre Monaghan as Vampire 1, Kate Minogue as Vampire 2 and Cathryn Brennan as Vampire 3.

Tom McDevitte, at only twenty-six, could be said to have written and presented the original Radio Éireann soap opera or serial. *Barney McCool of Coolaghey* started in 1938 and ran fortnightly for over a decade. McDevitte invented the imaginary character Barney McCool from the townland of Coolaghey near

Shakespeare's plays adapt well to radio with actors playing multiple roles in casts. Irish actor Thomas Studley (left) and director Hilton Edwards rehearse *Hamlet* for broadcast on Radio Éireann on 13 December 1964.

Strabane, Co. Tyrone. Every second Saturday, McDevitte cycled to Dundalk and took the train from there to Connolly Station to the Dublin studio. Through all the voices of the characters he invented and played, McDevitte told listeners live about all the exciting happenings and gossip in McCoole's village, which seemed to be the centre of the universe. Listeners got to know the characters well. Even though McDevitte moved to Belfast in 1945 and his train journey became longer, he continued with his serial for some years. His son Barry later met many people including Catholic Archbishop Sean Brady (later cardinal) who remembered being told to be quiet for fifteen minutes because 'wee Barney' was on the wireless.

The Dublin-based drama *The Foley Family* ran from 1947 until 1965 as an occasional series, usually consisting of six episodes at a time. The *Foley Family* signature music was 'Le Jet d'Eau' by Sydney Smith. One listener recalled how, as a boy growing up on a farm, once the piano music ended their dog used to get up and walk towards the door, knowing it was time to head out to bring in the cattle for milking. Another listener remarked how the Foleys were like the Ewings in *Dallas* in the 1980s, and how people would be all talk on the bus the next day about the latest exploits of Tom and Alice Foley; and remember this was at a time when nothing on radio could be heard except at the time it was broadcast and there was the immediacy of people having heard it at the same time as each other, or not at all. Another listener remembers when, having moved from the country to Dublin to work, they attended the Abbey for the first time when it was temporarily in the Queen's Theatre following the 1951 Abbey fire. He was enthralled to see what some of the actors from *The Foley Family* looked like, immediately recognising them on the stage by their voices from the radio.

The *Foley Family* writer was David Hayes, who also worked as an Aer Lingus executive. He had a natural gift for situation comedy and conversational technique, key to getting the attention of listeners. Hayes described his creation Alice Foley as the Elastoplast that stuck together the destruction that presented itself in every episode. Members of the Brennan theatrical family, Daphne Carroll (who played Alice Foley) and her real-life, eleven-year-old son Paul Brennan, who played Brendan Foley, featured in the mid-1960s cast. Daphne Carroll replaced Pegg Monahan, who had previously held the role of Alice Foley and was now acting in America. A contemporary article in the *RTÉ Guide* illustrates the interdependent ecosystem of Dublin's creative world and broadcasting. The effects man for *The Foley Family*, for example, was also the star vocalist-pianist of the popular Greenbeats showband and the Radio Éireann *Not So Green* comedy series. Others in the cast were architect Frank Purcell, journalist Jim Phelan and Telefís Éireann announcer David Timlin.

The Kennedys of Castleross was another huge hit with Radio Éireann listeners in the 1960s and early 1970s. Sponsored by

Fry's Chocolate, the fifteen-minute programme ran until 1975, when RTÉ was phasing in conventional radio spot advertising (ad breaks during and between programmes) in place of sponsored programmes at lunchtime, when it was broadcast. Nearly every radio set in the country was tuned in to their twists and turns. There were even stories of factories being forced to change their lunch hour so that workers could get home in time to hear it. It had many writers including David Hanly, later known as presenter of Radio 1's *Morning Ireland*. Hanly had already gained experience writing for RTÉ Television's fictional rural family, *The Riordans*. Like *The Riordans*, *The Kennedys of Castleross* covered contemporary agricultural life while keeping listeners in Ireland – as well as Britain – hooked with intriguing human drama. And after its departure from television, *The Riordans* was granted a brief period on radio too.

The cast in 1965 of *The Kennedys of Castleross* broadcasting live from Radio Éireann. From left to right: Máire Ní Dhomhnaill, Marie Kean, Vincent Dowling, unidentified, unidentified, Philip O'Flynn, Angela Newman, unknown, Aideen O'Kelly, Jim FitzGerald and producer William Styles.

Lee Dunne featured on the *RTÉ Guide* cover of 2 May 1975. He said that as the writer of the radio drama *Harbour Hotel* he wrote two episodes a day to provide the required five weekly episodes.

Lee Dunne's *Harbour Hotel* was the weekday lunchtime sound from 1975 to 1990. Hotel proprietress Kathryn O'Connell had returned from America to try to make a fist of running the place despite opposition from a local businessman who wanted to get hold of it for himself. *Convenience Corner*, also written by Dunne, was the final such lunchtime drama to be broadcast on RTÉ Radio.

RTÉ Radio won a coveted Prix Italia award for the first time in 1961 for Micheál Ó hAodha's adaptation of *The Weaver's Grave* from a short story by Galway writer Seamus O'Kelly. It is a deceptively simple telling of the arguments of two old men over

6 RTV Guide, December 1, 1961

Jubilation in Henry Street

Triumph for RE Rep actors

WHEN a telegram in Irish came to Radio Eireann last September with the good news that "The Weaver's Grave" had won the Italia Prize, not even Michéal O hAodha, who adapted the script and produced the programme could have been more jubilant than his colleagues in Henry Street.

The Italia Prize is not just any honour. It is the greatest radio honour to be awarded internationally. The Italia Prize was initiated in 1948 by the Italian Radio, which brought together the representatives of fourteen broadcasting organisations. Since then, the numbers of both prizes and of member organisations have grown. The competition is now truly world-wide.

Some years ago, I served on one of the Italia Prize juries in Taormina. I remember the September heat of Sicily burning the air in the spacious, elegant and shuttered room where the offerings were heard. I remember the exquisite courtesy of the Italians and the hospitality, the babel of languages around the long table, and the readiness of the Italian interpreter, who spoke English with a good Dublin accent.

The sharpest memory is, however, of the intense, mercilessly critical minds of the jury. They listened, argued, weighed and judged, as if it were a matter of life and death.

The Radio Eireann Repertory Company group who won the 1961 Italia Prize for their performance of *The Weaver's Grave*, with producer, Michael O hAodha. In front (l. to r.): Peggy Monahan, Michael O hAodha, Maura O'Sullivan. Second row: Eamonn Kelly, Brendan Cauldwell, Frank O'Dwyer, Tomas Studlai. At back: James Mac Hale (balance and control), Arthur O'Sullivan, Henry Comerford.

***The Weaver's Grave* adapted by Micheál Ó hAodha from the short story by Galway writer Seamus O'Kelly, won the coveted Prix Italia. The production overcame competition from broadcasters in countries including Japan, Australia, Canada and Britain, while the jury included radio drama professionals from France, Sweden, Germany and the Netherlands.**

a weaver's place of burial. Ó hAodha realised such an effective version of the original story by adapting the most vital descriptive passages into soliloquys for the characters. He played with the idea of them being so taken with self-description that they could, as it were, tell themselves to the listener in the third person.

The Weaver's Grave was broadcast on RTÉ Television and presented in the Peacock Theatre in recognition of the international acclaim it garnered, although it was generally agreed that the play's word imagery was best heard – a reminder of the unique art form of radio drama. The radio version went on to be produced in twenty other countries in different languages and is today available to hear, like so many more RTÉ Radio plays, on the Drama on One website.

Manchester-born actor, producer and writer Dan Treston won the Prix Italia award for Radio Éireann in 1965 for his play *Piano in the River*, which he'd been inspired to write after seeing a piano in the Liffey near his home. Talking to the *RTÉ Guide* of 22 October 1965 about the piano, Treston explained how, 'it seemed to me that while the piano might have been destroyed by the river it wasn't dead. Something of it must remain – some sound, for the piano is a sort of memory. Its meaning is different for each person who remembers it.' Performances were led by actors Thomas Studley and Deirdre O'Meara, and the piano player was Ian Henry.

Dan Treston's Prix Italia award-winning radio drama *Piano in the River* was first broadcast on Sunday 24 October 1965. The related cover drawing for that week's *RTV Guide* was by Alan Corsini.

On 10 September 1972, fifty years after the publication of *Ulysses*, RTÉ Radio broadcast a production of James Joyce's seldom-performed play *Exiles*. The broadcast was by way of an arrangement with BBC. Joyce's script was reworked by Harold Pinter, whose interpretation was then recreated for radio by Guy Vaesen. The Abbey Theatre would go on to produce their own version of *Exiles* for RTÉ Radio in 1974, which was later broadcast as part of the larger-scale celebration of Joyce's work during his birth centenary year of 1982.

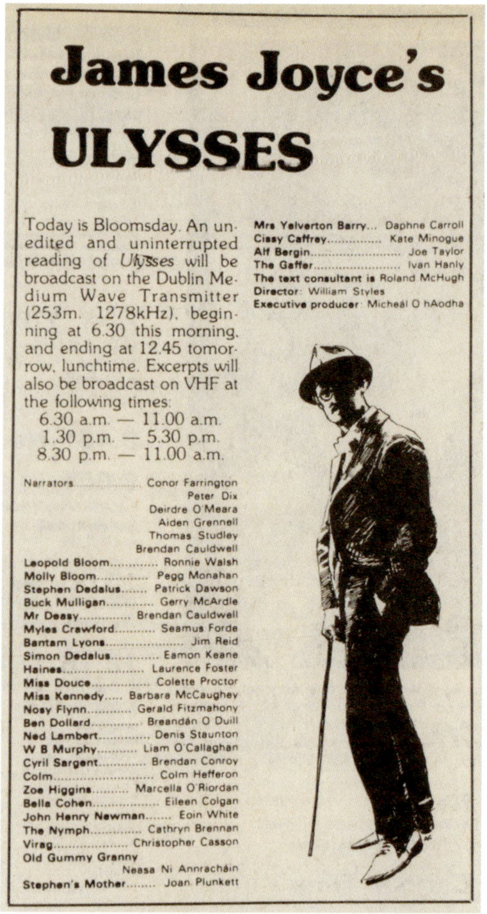

James Joyce's ULYSSES

Today is Bloomsday. An unedited and uninterrupted reading of *Ulysses* will be broadcast on the Dublin Medium Wave Transmitter (253m. 1278kHz), beginning at 6.30 this morning, and ending at 12.45 tomorrow, lunchtime. Excerpts will also be broadcast on VHF at the following times:

6.30 a.m. — 11.00 a.m.
1.30 p.m. — 5.30 p.m.
8.30 p.m. — 11.00 a.m.

Narrators .... Conor Farrington, Peter Dix, Deirdre O'Meara, Aiden Grennell, Thomas Studley, Brendan Cauldwell
Leopold Bloom .... Ronnie Walsh
Molly Bloom .... Pegg Monahan
Stephen Dedalus .... Patrick Dawson
Buck Mulligan .... Gerry McArdle
Mr Deasy .... Brendan Cauldwell
Myles Crawford .... Seamus Forde
Bantam Lyons .... Jim Reid
Simon Dedalus .... Eamon Keane
Haines .... Laurence Foster
Miss Douce .... Colette Proctor
Miss Kennedy .... Barbara McCaughey
Nosy Flynn .... Gerald Fitzmahony
Ben Dollard .... Breandán O Duill
Ned Lambert .... Denis Staunton
W B Murphy .... Liam O'Callaghan
Cyril Sargent .... Brendan Conroy
Colm .... Colm Hefferon
Zoe Higgins .... Marcella O'Riordan
Bella Cohen .... Eileen Colgan
John Henry Newman .... Eoin White
The Nymph .... Cathryn Brennan
Virag .... Christopher Casson
Old Gummy Granny .... Neasa Ni Annracháin
Stephen's Mother .... Joan Plunkett
Mrs Yelverton Barry .... Daphne Carroll
Cissy Caffrey .... Kate Minogue
Alf Bergin .... Joe Taylor
The Gaffer .... Ivan Hanly
The text consultant is Roland McHugh
Director: William Styles
Executive producer: Micheál O hAodha

(Left)
James Joyce and Ernest Hemingway shared the cover of the *RTÉ Guide* of 8 September 1972.

(Right)
The schedule in the RTÉ Guide for the now celebrated first broadcast of the complete text of Ulysses by James Joyce by the RÉP on Bloomsday 1982.

The definitive voiced adaptation of *Ulysses* is widely considered to be one of RTÉ Radio's finest productions, and certainly one of its best known. Its director, William Styles, was called to the office of P. J. O'Connor, then Head of Drama and Variety. O'Connor wanted RTÉ to broadcast a complete reading of *Ulysses* for the centenary. Styles didn't let the fact that he had not actually *read* the book get in the way of such an ambitious undertaking and went on to study the text assisted by Joycean scholar Roland McHugh. By the time Styles had finished reading *Ulysses* the project had developed in his mind from a broadcast to a fully dramatised reading using the RÉP. Nearly a year later – and in time to broadcast the production on Bloomsday 1982 – Styles and the actors emerged from Studio 9 with a 29-hour-and-45-minute radio programme, one of the longest single dramas on radio ever made. Styles complimented the actors by declaring that they were nothing like the characters they portrayed in spirit, character or years.

Styles tried to stay as faithful to Joyce's text as possible while acknowledging challenges such as creating songs that only existed in Joyce's imagination. How, for example, to dramatise a note sung at the Ormond Hotel which lasted over two and a half pages in the text of the book and five minutes on radio? Since 1982 this production of *Ulysses* has set the standard for every adaptation of an already existing text in terms of quality and ambition. It continues to be popular listening on the RTÉ Radio Drama website and on CD, as noted in this 21 October 2023 letter in *The Irish Times*:

> Sir, – There is a simple solution available for those readers who are exhausted after reading some books that have abandoned quotation marks around dialogue. Listen to them as audiobooks. It is not a recent innovation. I managed to complete Ulysses with the benefit of the 1982 RTÉ radio drama production on compact disc. – Yours, etc.

The RÉP recording *Ulysses* in the drama studio in the Radio Centre.

Marie Kean, Bosco Hogan, Dermot Crowley and Maxine O'Reilly rehearsing *Big Maggie* by John B. Keane that was broadcast on 5 July 1970.

Sean O'Casey's *The Shadow of a Gunman* was broadcast on 6 September 1976 in *The Monday Play* slot. From left: producer John Lynch, unidentified, Niall Buggy, Donal McCann, Jim Reid, Daphne Carroll, Colette Proctor, unknown and Ginette Waddell.

Cast members from St John Ervine's *Boyd's Shop,* the first in a season of Irish comedies in June 1980. From left: J.G. Devlin, Catherine Gibson, Ivan Hanly, Derek Lord, Kate Minogue, Cathryn Brennan and Barbara McCaughey.

Recording the 1980 production of Donagh MacDonagh's verse play *Happy as Larry* adapted and produced by Seán Ó Briain.

Over the years attempts were made to find new plays for radio and occasional competitions were held to encourage scripts being submitted. In 1982 RTÉ Radio Drama established the annual scriptwriting award that continues today, named in honour of P. J. O'Connor. In 2004 RTÉ began its close involvement with the All-Ireland Drama Festival, recognising the centrality of theatre-making in Irish life.

Fishamble and lyric fm collaborated on a project that was a first for both in the early 2000s. Jim Culleton leads Ireland's new play company and directed his first play for radio with lyric fm's Eoin Brady. They brought listeners four new, water-themed and music-related plays: *Handel's Crossing* by Joseph O'Connor was inspired by Handel's *Messiah*; Gavin Kostick's *The Medusa* by Berlioz's *Benvenuto Cellini*; Michael West's *Chaste Diana* by Rossini's *The Barber of Seville*; and Stella Feehily's *Sweet Bitter* by Massenet's *Werther*. In 2026 a short drama series called *Jó Johnny Joe* was jointly broadcast on RnaG and Radio 1, combining music, comedy and grief in a moving, tension-filled story. Its writer, director and producer is Darach Mac Con Iomaire. *Jó Johnny Joe* has been made available to enjoy subtitled in English and in Irish on the Drama on One website.

The Abbey and RTÉ Radio have been interwoven for decades. Indeed, Willie Fay – one of the two Fay brothers who helped found the National Theatre and who played a large part in building up the success of the Abbey – was very nearly appointed first director of 2RN. One can only wonder what direction Irish broadcasting might have taken had this happened. What might listeners have heard under his leadership?

In 1986 Tom Murphy's *The Gigli Concert* was Radio 1's Play of the Week in the Tuesday evening drama slot of the time. The play had premiered in the Abbey in 1983. On seeing it, radio-drama producer John P. Kelly, as he put it, couldn't 'get away from feeling that here was something superb that could have been written for radio. You have two men in a room, interacting, teasing, intellectualising, crying, changing – and the music, the glorious emotional music of Gigli.' On Easter Sunday 1999 Marina Carr's play *By the Bog of Cats* had its radio premiere. Already awarded ESB Best New Play of 1998, audiences had packed the Abbey to see it. RTÉ's Anne Walsh had proposed a radio broadcast to

Abbey director Patrick Mason, who had also directed the play so that, as she put it, 'listeners throughout the country can hear the plays that are being talked about that they might have missed in the theatre, while theatre companies and radio itself, obviously, will benefit by having this hugely valuable [sound] archive.'

Her only playboy: Karen Ardiff as Pegeen Mike and Pat Kinevane as the man killed his da in the Synge Season's *Playboy of the Western World:* 8.02pm

Karen Ardiff as Pegeen Mike and Pat Kinevane as the man who killed his da in the production of *The Playboy of the Western World* that was part of the 1999 John Millington Synge RTÉ Radio Drama Season captured in the *RTÉ Guide*.

The year 1999 also marked the ninetieth anniversary of the death of John Millington Synge; RTÉ actor and producer Kate Minogue seized on the opportunity to bring listeners his plays alongside commissioned introductory talks, which chimes nicely with the tradition of a written programme note setting the scene for theatre audiences. Declan Kiberd, Marina Carr and Frank McGuiness gave the talks. Minogue realised the importance of producing performances by the finest contemporary actors for present-day audiences – and preserving them for future listeners – when she discovered the 1948 *Riders to the Sea* radio recording in the RTÉ Archives, featuring Máire O'Neill (Molly Allgood, Synge's fiancée).

Radio and the spoken voice have a capacity to reduce the distance between us and what we think is long ago. When we hear the human voice animating direct experience, we can hear history as new. What follows somehow seems apt to include in 2026, the centenary of Irish broadcasting and of the first production of *The Plough and the Stars*. In 1976, on RTÉ Radio's arts programme *Appraisal*, Gabriel Fallon – who had played the part of Captain Brennan in the first production of Sean O'Casey's *The Plough and the Stars* in the Abbey – recalled the famous riots it provoked fifty years previously:

> I remember saying to myself, you know, history is being made here. When you go home, take notes, which I did. And I said, this is a lamentable state of affairs, Mr Yeats. And he said, Fallon I have just sent for the police and this time it will be their own police. Then I began to see, of course, that there was a political motive as well as any other motive behind what was happening. So, he gave a signal to the stage manager to raise the curtain when he raised his hand, and the curtain was raised. And Yeats paused. Really, he was a marvellous actor. Waited for them, the boos, just barely to die down, lifted his hand in that familiar fashion and said, You have disgraced yourselves again ...

J. M. Synge's *In the Shadow of the Glen* in rehearsal in the Radio Centre with from left Mick Lally, Declan Conlon, director Garry Hynes, Marie Mullen and Ray McBride. It was broadcast in 1999 for the Synge Season.

Ciarán Hinds and director and playwright Patrick Mason in the Radio Centre in 2012 recording *The Dreaming of Roger Casement*.

Recording Eimear McBride's *Mouthpieces*. From left: Eimear McBride, actress Aoife Duffin, series producer Kevin Reynolds, sound engineer Gar Duffy and producer Kevin Brew.

In 2024 the Abbey Theatre in collaboration with RTÉ presented *Spreading the News*, a performance of new writing, as well as from plays deeply associated with the Abbey's story. Some of the participating actors had also played the same parts in previous RTÉ productions of the plays. *Spreading the News* was broadcast on Radio 1 on 27 December 2024, the Abbey's 120th birthday, in turn multiplying its original capacity audience in the theatre.

On 15 January 2026, to mark the centenary of the opening in 1926 of 2RN, RTÉ broadcast a day of celebratory programmes hosted by Radio 1, 2FM and lyric fm from the GPO, where the station had broadcast for nearly fifty years. Among the guests to contribute to Radio 1's weekday arts programme *Arena* that day from the GPO were Mark O'Brien, managing director of the Abbey and celebrated screen actor Bríd Ní Neachtain, also long associated with the national theatre. O'Brien mentioned to presenter Rick O'Shea that he couldn't help noticing how close the GPO building and the Abbey Theatre were to one another when he walked over from the theatre for the broadcast, as though tracing the footsteps of so many actors who had often done the same in the past. O'Brien also remembered the contributions to the Abbey made by Seán Rocks, the long-standing presenter of *Arena* until his untimely death in 2025. As an actor in the late 1980s and 1990s, Rocks had key roles in the premieres of two major plays there: Tom Murphy's *The Wake* and Marina Carr's *Portia Coughlan*.

O'Brien said the Abbey and RTÉ shared a common purpose. How, just as major social issues emerged on the likes of the *Gay Byrne* and *Marian Finucane* radio shows, it is noticeable that often what playwrights write about come in time to be the next major social changes or even the subjects of future referendums in the country. And how in allowing these subjects to be discussed and played out on radio and in facilitating artists to explore them on stage, RTÉ and the Abbey haven't been without their controversies. That too, is what makes them important. That they are part of the nation's dreaming, the people's dreaming about what the nation can be, and as such, are institutions worth fighting for.

It seems somehow only natural how over time radio has become embodied in Irish plays, sometimes as the source of a narrative, sometimes as a character. Bairbre Ní Chaoimh and Yvonne Quinn's play *This Was Your Life* was inspired by writer Maura Laverty. Laverty, after all, was a playwright, scriptwriter, novelist and journalist, as well as host of the Radio Éireann ESB-sponsored programme. Their stage play was adapted for radio, with Ní Chaoimh – herself an experienced writer, broadcaster and actor – once more in the role of Maura Laverty. It was directed for Drama on One by RTÉ's Goretti Slavin.

And of course, there is Marconi, the radio in the kitchen of the five Mundy sisters. The radio releases them, even if just temporarily, out of themselves so magnificently, so heartbreakingly in Brian Friel's 1990 play *Dancing at Lughnasa*. Just as his narrator recalls: 'I remember my first delight, indeed my awe, at the sheer magic of that radio ... the kitchen throbbing with the beat of Irish dance music beamed to us all the way from Dublin.'

Woven into Éilis Ní Dhuibhne's 1995 play *Dún na mBan Trí Thine* (The Women's Fort is on Fire) is the ritual of broadcasting the local death notices on RnaG. Ní Dhuibhne's play powerfully manifests the observation by poet Seán Ó Riordáin, who said that after the establishment of the station you could not be truly dead until it was heard on the radio. And of course, there is Martin McDonagh's *The Beauty Queen of Leenane* with the radio, and what we hear come from it on stage – Delia Murphy singing 'The Spinning Wheel' – highlighting the deadening life of the play's characters being played out in front of us.

(Opposite page) Caitríona Ní Mhurchú's bilingual play *Eating Seals and Seagull's Eggs* premiered at the Dublin Theatre Fringe. It considers loneliness, the female voice, state-making, censorship and Peig Sayers. Goretti Slavin directed the play for Drama on One in 2019.

(Opposite page) Irish writer Maura Laverty is the subject of *This Was My Life* by Yvonne Quinn and Bairbre Ní Chaoimh, who also played the part of Laverty. It was produced for Drama on One in 2021 under the direction of Gorretti Slavin.

POETRY AND PROSE

David Hanly's voice introduced his new Saturday poetry series *The Enchanted Way* in 2001 as a request programme, with a couple of his own in there for good measure. 'Advent' by Patrick Kavanagh was the choice of Rita Keyes of Rathfarnham in Dublin, who said she had attended a retreat where the poem was analysed line by line. Hanly said he didn't intend to go as far as that because he remembered just such a dissection of another poem, 'Soft Ease or Bard of Passion and of Mirth' by John Keats, by two American academics in Scariff many years before. Hanly's wry aside on its own was worth the listen. The choice of poems wasn't bad either. Doubtless I remember the programme so vividly because of the intensity of my own situation at the time. I had been brought into Holles Street earlier that day, two weeks overdue. Our daughter, born four hours later, has a long-standing affection for William Butler Yeats, who also figured on *The Enchanted Way* that evening.

Poetry featured on 2RN from its first month on air. Frank Fay broadcast a selection in January 1926 and continued to do so for some time. But as ever, he couldn't please everyone and thus the *Sunday Independent* of 16 November 1930 declared: 'Please ... stop these dreadful Poetry Recitals ... I never knew I disliked poetry until they put it through on the wireless.'

Dublin poet Austin Clarke presented a poetry programme called *Anthology* from 1942 to 1968, which Seamus Heaney would fondly recall. Numerous poets, fiction writers and broadcasters passed each other the baton of presenting what has been a continuous presence of poetry programmes in English and Irish across the century.

From poets such as Brendan Kennelly, Eavan Boland, Seamus Heaney, Mary O'Donnell, Pat Boran, Vona Groarke and Rachel Hegarty to broadcasters including Olivia O'Leary and Rick O'Shea, each series took a different approach and explored a distinct focus.

Austin Clarke, pictured in 1948. His long-running programme was hugely influential.

Máirtín Ó Díreáin in 1971, whose poetry was regularly featured on RTÉ.

Paul Durcan in full flow at the live radio and television broadcast of the short-lived Bank of Ireland Arts Show Awards.

Michael Longley with presenter of Radio 1's *Poetry Programme*, Olivia O'Leary.

Young Writing, presented by Eavan Boland, was a new concept on radio in 1969, targeting those between 'fifteen and twenty-five' who were interested in writing. Reporting on the associated Young Writers' Seminar in the Gresham Hotel, Dublin, the *RTÉ Guide* remarked that:

> the questions reflected a youthful and perhaps idealistic notion of what a writer's life is all about. There was an overriding concern with attitudes, and approach; the problems of getting started as a writer ... one's sanity through the first few years of probable disappointment – all seemed very low on the teenagers' list of priorities. Time enough to worry about such niggling trivia.

Award-winning writers Paul Lynch and Christine Dwyer Hickey shared their own associations with radio and the part it played in their becoming fiction writers when they contributed to *Arena* live from the GPO on 15 January 2026.

In 1993 Lynch and his brother were metalheads, and proud of it. On Sunday nights they listened to *The Metal Show* on 2FM, which introduced them to a range of music they had no access to. It was still a time when people recorded things off the radio onto cassettes. Lynch bought an old Vox guitar and his brother a tuneless drum kit with cracked cymbals and they formed a band. Having managed to get a demo together they sent it to the show's demo slot. When it got played on the programme, Lynch said it felt like they had just taken over 2FM from Carndonagh in Donegal. He learned that you just need to have some imagination, guts, vision and a lot of luck and that doors can open for you: 'This is the thing about Ireland. Everybody knows somebody who has had their moment on the radio.'

***Arena* Presenter Rick O'Shea talking with from left film director Colm Báiread, and writers Christine Dwyer Hickey and Paul Lynch on the programme that came live from the GPO on 15 January 2026.**

As a teenager Christine Dwyer Hickey spent months in solitary confinement in Cherry Orchard hospital, seriously ill with viral pneumonia. She had a little radio and used to listen to Gay Byrne and became interested in other people's lives. He had a blind man on the show. Women told Gay how lonely they were. A priest back from South America talked about drinking his own urine. It opened her mind to worlds outside of her own and it planted the idea of creating her own stories in time.

James Plunkett speaking to presenter Eavan Boland at the Young Writers' Seminar in 1969.

Edna O'Brien in conversation with Mike Murphy for his *Reading the Future* series in 2020.

The RTÉ Short Story Competition winner for 2025, Lynda McCarthy, being warmly applauded as she collects her award.

Sinéad Gleeson presented *The Book Show* on Radio 1 and has herself gone to become an editor and writer.

FROM THE PAGE

CHAPTER 4

The RTÉ Short Story Competition is for new stories written for broadcast on radio. L to R: Stephen O'Reilly, second prize for 'Divination'; Emer O'Toole, third prize for 'The Other Órla'; *Arena* presenter Seán Rocks; and Mattie Brennan, first prize for 'Dead Bait' in 2024.

Our favourite poet, Seamus Heaney, famously wrote about digging with his pen. Having produced thousands of radio essays as producer for several years of *Sunday Miscellany* and other writing-based programmes, I know the power of the pen and the impression it can make on listeners.

Since its earliest days as 2RN, RTÉ Radio has welcomed submissions of writing for undramatised broadcast from journalists, writers and academics as well as the general listener. Such writing can vary in length and purpose: informational, scholarly, entertaining or a mix of all three. In many ways, these radio essays or 'talks' are the building blocks of radio, enduring and adaptable, holding attention through the strength of a single voice.

The earliest 2RN broadcasts – other than musical and dramatic performances and live reportage – were almost entirely scripted. Journalist J. B. Hall wrote in the *Irish Independent* of 21 November 1927 about what went on inside the heads of the earliest contributors to radio:

> It is doubtful if a more unique experience can well be conceived than that of the individual who makes his or her first plunge, as it were, into the billows of broadcasting ... To do a 'talk' for fifteen minutes, a lecture, or humorous or pathetic recitation, may to the uninitiated seem simplicity itself. Pleasant as the anticipation may be to the performer from a distance, yet when

the date and hour draw near, the feeling in many cases assumes a remote resemblance to that of a person bent on having a tooth extracted. He may even pause in doubt as when he reaches the dentist's door. But the sensation quickly vanishes and is replaced by one of weird fascination in addressing an unseen audience ... There are two besetting sins of which the beginner is generally guilty – one is that he talks too loudly, and the other that he will rattle on at the rate of a hunt until the eloquent look or gesture, or maybe written admonition, curbs his wild career, and he settles down to that moderation and unaffected deliberation which are the golden rules of Wireless when the Announcer proclaims 'Dublin is Calling'. And it is time to begin.

These days talks are generally recorded before broadcast. Almost a century after Hall I like to think that – similar to other colleagues – I have reassured contributors recording for the first time that it is simply a performance of their own words and that it is natural to do a re-take and reasonable that they would leave something for the sound engineer and producer to work with before the listener hears their piece on air. Nevertheless, it is important to harness the focus and energy that the red light on the microphone signals.

RTÉ has published anthologies out of *Sunday Miscellany* since 1975.

As a child, I remember being shushed to be quiet during *Sunday Miscellany*. It wasn't personal or special to our kitchen (or car), that shushing. Others told me it had been the same for them, this gesturing by parents to either join in the listening, or move along out of the room, or to be silent on a drive, especially during the contribution of a particular voice or two. It wasn't that they were cross. Not at all. It was, rather, for the pleasure of listening: they didn't want distraction when the programme was on the radio.

Maybe it was because something was broadcast just once. Or perhaps a snippet of it made it to *Playback*, at the weekend. We could hardly have imagined that, years later, the programme – or any segment of it – could be listened to live or anytime you fancied. Perhaps it also had to do with the programme being broadcast on Sunday morning and people generally not needing to be somewhere. Open to being transported elsewhere. Indeed, the Sunday morning aspect of it reminds me of how *Sunday Miscellany* used to be described by some as mass for atheists!

Listening to its archives yields treasures from among its contributors, such as Val Mulkerns, John Jordan, Maeve Binchy, Seán Ó Faoláin, Mae Leonard, Cyril Kelly, Paul Howard, Denise Blake, Tim Carey and Vona Groarke. Each decade added more voices. Established solid ballast to navigate its ongoing story. To check in on itself. To know it had a lineage to take strength from but also to question and challenge and make new again. It seems to me that a constant appeal of its contributors is their own pleasure extended to us in telling us what they have to say and in how to put it in their own words.

Indeed, in a short piece broadcast in 1975, Maurice Craig talks about the origins of the art of short writing for radio and a man called A. J. Alan. Craig declared Alan to have been 'the man who invented the art of talking on the wireless ... as though he were talking to someone in front of him and not to a microphone'. He goes on to tell how Alan collected a selection of his writing for BBC Radio in a book published in 1928 called *Good Evening, Everyone*. He draws attention to Alan's self-consciousness of what made him a success, how 'after taking endless trouble to sound natural, he [Alan] is afraid, however mildly, of looking too natural in print. It is, or was, not the done thing to appear in public in one's shirt sleeves, so to speak.'

Ronnie Walsh, then presenter of *Sunday Miscellany*, featured on the *RTÉ Guide* cover to mark the publication of the second *Sunday Miscellany* anthology in 1976.

Although Craig considers many of his turns of phrase and references to be dated and the content light, he thinks Alan deserves to be remembered for his radical breakaway into sounding natural on air, when until then, as Craig wryly put it, radio could be seen as 'simply a medium for reading leading articles to people who are too lazy to read them for themselves'. Alan reportedly took months to mull over his material and how to present it to the listener. It is something the most effective *Sunday Miscellany* contributors still do.

I like to think that 2RN's first director, Séamus Clandillon, may have been introduced to or even witnessed A. J. Alan in action in the BBC. In his formal letter of appointment from the secretary of the minister for posts and telegraphs, J. J. Walsh, in November 1925, Clandillon was encouraged to 'proceed at once to London to get an insight into the practical working and control of broadcasting'. The letter went so far as to tell Clandillon that, on the boat from Dún Laoghaire the next day, T. J. Monaghan, engineer in chief of the department, who also had responsibilities for the new station, would bring him up to date on matters on the boat.

It may have been due to Séamus Clandillon's absence that the opening night of the station was primarily music-based – the station's music director, Dr Vincent O'Brien, having been appointed earlier in 1925. All the same it was Clandillon who corresponded with the musicians about their engagement and music they would perform on the opening night. He gave the excuse that, because of the Christmas holidays, it had become impossible to arrange for a suitable series of spoken contributions by well-known people before the station's opening. Considering the excitement palpable in letters and cards between the station and its calibre musical contributors in advance of opening night, what Clandillon mentioned seems a little lame by way of explanation for no speech being included in the inaugural broadcast.

It was only after the station opened that Séamus Clandillon put an announcement in the newspapers welcoming 'suitable short talks and lectures on subjects of national or literary interest, or on such subjects as agriculture, dairying, horticulture, and ... from people capable of giving short, interesting talks in Irish on subjects of general interest'. He could offer only small fees to contributors, which limited what was offered to listeners. (The station was competing for financing with its own department as well as with multiple government departments in the throes of new State-building decision-making including power, health, education, housing and defence services.)

Máiréad Ní Ghráda – who worked in the station between 1927 and 1935, continuing as an occasional freelance after that – recalled, on the fortieth anniversary of the station in 1966, how staff and contributors to its programmes worked hard and weren't in it for the money. She spoke of a sense of excitement – like having a new toy, she explained – and how those involved were up for learning from trial and error. Somewhere in between the adventurous spirit she remembered and Clandillon's limited budget, the station forged a way into the daily soundscape of Irish life.

The first months of 1926 included broadcasts of talks in Irish on poet Seathrún Céitinn, *Gaeil na Seachtú hAoise Déag* and *An Seana-Shaol* (the old life) by S. Ó Duirinne, a talk on amateur photography by H. L. McNally and by H. N. Kirwan on the Huguenots in Ireland. Gerald Ormsby – or G. O. Sherrard, as he was known on air – taught at the Albert Agricultural College in Glasnevin, Co. Dublin. He gave his first talk on gardening in 1926 and continued into the 1950s. He also had a gardening column in *The Irish Times*. RTÉ Archives acetate recordings include Sherrard talking in 1946 about sowing spring onions and choosing autumn flowers for the garden:

> the dahlia likes to be treated like the torched lily, and a plant of one of the larger varieties, given a bed to itself, treated like a flowering shrub. But it would need careful staking owing to the weight of the head and brittleness of the shoot. It has come to the front very much as a bedding plant in recent times ... some people don't like dahlias in the herbaceous border, but I think they provide reds and pinks which are scarce there at this time of year ...

Sherrard was reluctant to plant the perennial sunflower in his own small border on account of its spreading capacity and was surprised to see it in his sister-in-law's border in Cork. A bit of horticultural time travel to hear him.

The new state recognised the usefulness of the station to distribute information to the public even if there were only some thousands of radio listeners in 1926. Sometimes reports of even the shortest radio talks appeared in the newspapers so that those who didn't hear the broadcast could read about what was said, and those who did could mull over what they had heard. On 16 and 17 April 1926, as well as on 18 April, the day of the census itself, Stanley Lyon, director, statistics branch, department of industry and commerce, gave talks explaining the census. Lyon also provided similar talks in 1936 and 1946. Public-service-related talks on health and safety were sometimes broadcast in Irish as well as in English. In 1943, towards the end of the Second World War, D. J. Giltinan gave a series of seven instructive talks on air-raid precautions. He claimed £21 for writing the scripts but Radio Éireann declared that they had been provided free of cost to the station. The district courts ruled that, since there was no written contract, Giltinan's claim was dismissed: a lesson for both sides.

It was not long after he became station director in 1935 that T. J. Kiernan persuaded the department of posts and telegraphs to provide funds to purchase equipment that allowed for short-length acetate recording. Kiernan's wife was the popular singer Delia Murphy. He understood the value of recording. The station purchased tape-recording equipment in the late 1940s, which

G. O. Sherrard presented a weekly talk on gardening on Radio Éireann from 1926 until the 1950s. He was the subject of the Portrait Gallery in *The Irish Times* on 12 December 1953, where he contributed a weekly gardening column.

PORTRAIT GALLERY

TO-DAY is Saturday. All over Ireland—even though it is December—hundreds of men will be looking out the window, eyeing the sky, wondering whether the afternoon will be fine enough for a couple of hours digging the vegetable plot or planting the new rose-bushes. These are the amateur gardeners: bank clerks, shop assistants, garage owners, income tax inspectors by profession, but in their spare hours followers of the oldest and the most universal hobby in the world.

G. O. Sherrard

To almost all of these people, G. O. Sherrard ranks as Ireland's Number One gardener. Weekly, for 27 years, they have listened to his talks from Radio Eireann—a record that makes him probably the nation's senior broadcaster; for over ten years they have read his advice in the columns of this newspaper. Those who attend the horticultural shows may be familiar with his tall, military figure, moving with the slow dignity that seems appropriate to a man who has made his career among the vegetable kingdom.

The military bearing is more than an accident. Gerald Ormsby Sherrard was born in Cork, where his father farmed near Douglas village; but he was early sent to England to a minor public-school, where he was prepared, along with most of the other pupils, for a career in the Army. He was commissioned in the Royal Artillery, and posted to Gibraltar. There—fortunately, as it turned out—he fell ill with Mediterranean fever. On a long sick leave, he revived a boyhood interest in gardening: and finally a course of the writings of Horace Plunkett decided him to resign his commission and take up horticulture seriously. He left the Army in 1902, returning to it only to serve in the 1914-18 war.

Back in Ireland in 1903, young Sherrard trained at the Albert Agricultural College—the oldest institution of its kind in these islands — and the College of Science. In 1907, he was awarded a scholarship under the Development Commission, to go to the famous John Innes Institute, in England. After the interruption of the war, he returned to the John Innes Institute, but soon afterwards was appointed the first lecturer in horticulture at the College of Science.

To-day, he is still Professor of Horticulture in the National University; most of his work is done at the Albert College. He has trained almost all Ireland's horticultural instructors. In addition, he has given his time lavishly to the societies: he is at present President of the Royal Horticultural Society of Ireland and also of the professional body, the Horticultural Education Association. His own work has been chiefly on the breeding of Brussels sprouts; showing a basket of them, he will say, with his characteristic slow smile: "These are the finest sprouts in the world." In fruit, his chief interest is apples, and he sees a big future for co-operative packing and marketing on the lines of the Dungarvan Co-op.

At home in Pembroke Park, he is one of thousands of suburban gardeners. "I don't get much time nowadays," he says, "but I have to keep the front garden looking decent." His own flower beds, like a dentist's teeth, are the standard by which people judge him. He has a small piece of ground, too, on Bantry Bay, where he delights in growing rhododendrons.

The Sherrards are, of course, a "planter" family. They came first to Derry, in the Elizabethan plantation. Three of them distinguished themselves at the siege of Derry—two among the Prentice Boys, one as officer of the Town Guard. If their descendant takes a modest pride in this family history, not the most fervent Gael would grudge it to him. G. O. Sherrard has given a new meaning to the word "planter."

allowed for longer recording and easier editing. As other forms of programming developed, radio talks became a choice for more considered, uninterrupted expression and could be recorded and edited before broadcast.

The 1950s saw a further innovation when Maurice Gorham became director in 1953 and introduced the Thomas Davis Lectures. Their ambition was later outlined in the 1955 *Radio Éireann Yearbook* by Trinity College Dublin historian Dr T. W. Moody. Moody was, at the time, a member of Comhairle Radio Éireann. The Comhairle, established in 1951 by Erskine Childers, then minister responsible for broadcasting, was the first formal step towards the station's independence from the department of posts and telegraphs. Moody wrote that 'a national broadcasting service, if it is to be national in the highest sense, must cater for a wide range of interests and needs, for listeners in relatively small numbers as well as for the tastes of the majority'. He drew attention to the strides being made in the provision of music on Radio Éireann. The first classical music 'promenade concerts' had been introduced to listeners in early 1953 and the traditional music offering was greatly enhanced by the gathering of music

***Tuairisc an Bhollscaire* (Announcer's report) of 16 April 1926 announcing a Census Talk on 16 April 1926.**

TUAIRISC AN BHOLLSCAIRE.
(ANNOUNCER'S REPORT.)

Programme as Broadcast.

STUDIO CONCERT. 16th April, '26

Denmark Street, Dublin.

Time. Duration.	Artist.	Item.	Composer.Publisher
p.m. 7.29.		Tuning Note.	
7.36. 14 mins.	F.E.Stephens.	The Growth of Greece.	
7.45.	Cathal McGarvey.	Humourous Monologue.	
8. 0.117 "	London Relay.	La Traviata.	Verdi.
9.59.		Weather Forecast.	
10.0.	Announcement re advertising Irish Industries.		
10.1.	do.	re Census talks.	
10.1½.	do.	informing Banbridge listeners of Pipe item on Tuesday.	
10.2. 24 "	Station Orchestra.	Unfinished Symphony.	Schubert.Lafleur.
10.27. 2 "	do.	Dances. Gavotte.	Cyril Scott. Boosey.

Deireadh Gno.

from across the country for broadcast by the mobile recording unit. Something equivalent was called for to offer academic scholarship with an Irish slant to listeners in English and in Irish. (Remember that in Ireland those who attended university in the 1950s were a privileged fraction of the population and that free secondary education was not introduced until 1967.) The Thomas Davis Lectures, whose first producer was Francis MacManus, were given prime airtime on Sunday evenings and attracted high listenership.

Moody referred to the BBC's Third Programme idea but knew that such an option was not financially viable for Radio Éireann. The Third Programme was established in 1946 as BBC's third channel. It was part of BBC's post-Second World War reorganisation to offer a choice of listening with a deliberate focus on classical music, drama, literature and discussion. The BBC director general, Sir William Haley, said 'its whole content will be directed to an audience that is not of one class but that is perceptive and intelligent'. (At the time, Radio Éireann's nod to listener choice was limited to it moving between the stronger or weaker wavelengths of its transmitters for different sports.)

The idea of a 'university of the air' was a major innovation for a single-station radio service like Radio Éireann to take on. Comhairle Radio Éireann agreed on the title of the Thomas Davis Lectures. Davis, a nineteenth-century Irish Protestant nationalist and poet, believed in the phrase 'educate that we may be free' and in the power of the printed word. Academics had not been drawn to the poor rates previously offered to present talks on the station. The half-hour lectures paid better and, from the beginning, were published in accessible book form: an important element for academics to extend their publication and for the general listener to spend time with them, especially before repeat broadcasts or listener recordings became available. Programme websites – not to mention podcast editions of programmes – were far in the future. The books were a nod to cross-media thinking in action in mid-twentieth century Ireland. (In time, more than a few of us in University College Galway were most thankful for the book of the Thomas Davis Lectures entitled *Morality and the Law* edited by its radio series editor Desmond M. Clarke. As the chestnut trees came into bloom and one of our first-year law

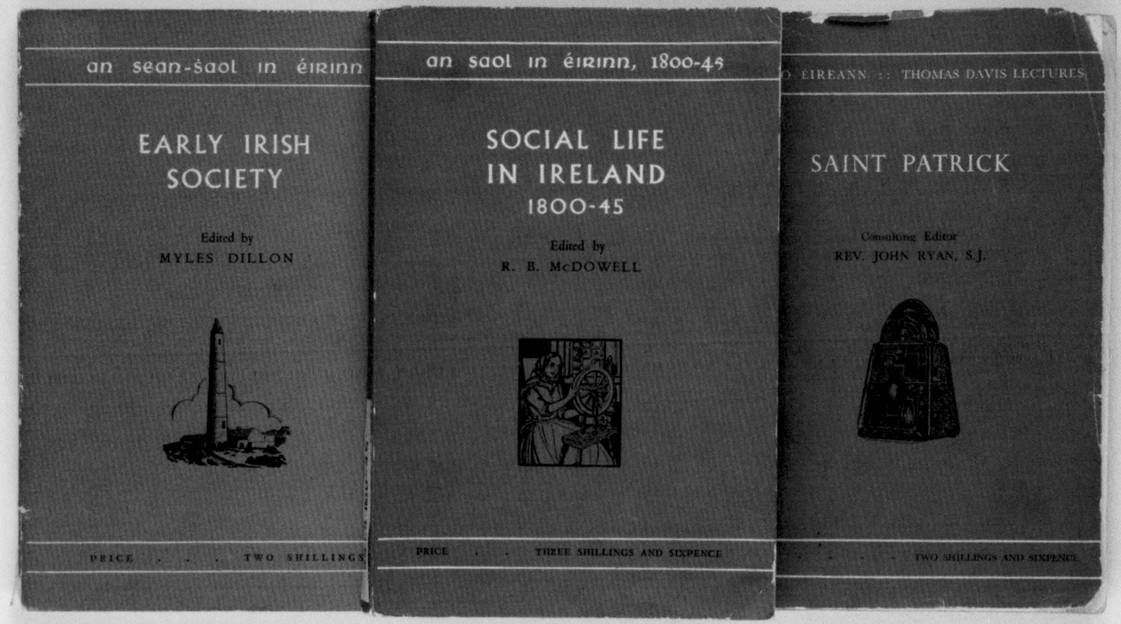

Published Thomas Davis Lectures.

exams loomed, I'm sure it was those talks, true to their original intentions to provide scholarship in a digestible form, that got us the exam!)

The first series consisted of six lectures under the general title of 'Early Irish Society' under the direction of Professor Myles Dillon of the Celtic Studies School at Dublin's Institute for Advanced Studies. His talk 'The Irish Language' was broadcast on 27 September 1953. Dillon's colleagues in the Institute for Advanced Studies provided the other talks in the series: David Greene on early Irish literature, M. A. O'Brien on Irish-origin legends, D. A. Binchy on secular institutions, James Carney on the impact of Christianity. David Greene also gave the final talk on early Irish society.

They were immediately followed by three unconnected weekly lectures by R. B. McDowell on Robert Emmett, Arthur Aston Luce on George Berkeley and Seán P. Ó Ríordáin on Tara, Co. Meath. Ó Ríordáin had led archaeological excavations at the ancient site of Tara and the inclusion of his illustrations of the digs in the subsequent publication of his talk only added to its value for the reader.

If there had been any doubt about whether the project would be attractive for listeners, the reaction to them proved otherwise. Radio Éireann was contacted by telephone and letters and even some called in person looking for copies of the text of what they had heard on air (the station had not learned yet the importance of a publication associated with a series being available to be purchased by the public at the time of the broadcast). Director of the station Maurice Gorham remarked on how previously sceptical academics and scholars who had 'declared "would you believe it, I was asked to do a broadcast by those fellows in the G.P.O.", were now complaining "It's a queer thing I haven't been asked to give a Thomas Davis Lecture."' Giving a Thomas Davis Lecture gained a status, perhaps especially when it began to take on subjects that were closer to the present, such as the 1957 series 'The Shaping of Modern Ireland', covering aspects of the country in the twentieth century.

The United Nations responded to the fact that the growth rate of the world's population was at 2 per cent when it proclaimed 1974 World Population Year. The designation inspired Michael Littleton, then producer of the Thomas Davis Lectures, to commission a series on the subject, illustrating the gradually more contemporary relevance of the choice of subjects of the lectures being broadcast. Professor Brendan Walsh of the Economic and Social Research Institute and Dr Dermot McAleese of the department of economics, Trinity College Dublin, led the series. All very well and admirable, a great subject choice, but not a single woman gave any of the eleven talks. There was a way to go in the balance of voices being represented, being heard.

A 1997 series showed improvement. Commissioned to mark her birth centenary, the subject was Kate O'Brien. It is poignant to hear Eavan Boland articulate what a present generation can gift those who come after them as she told the listeners 'the living memory of a writer is one of the real gifts which one generation can offer to the next. It's the gift the present makes to the future. After all, that is the way we add substance to style and accuracy to our reading.'

These days it is perhaps Boland's gradual realisation of how her appreciation of Kate O'Brien grew from admiration to professional influence, and the critical difference between the

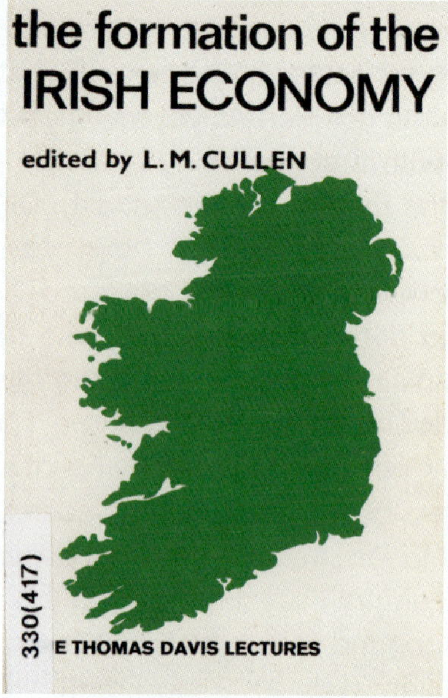

A selection of published Thomas Davis Lectures.

two of them, which strikes most. What sticks with you as a listener can take time to sink in. Boland related how she had kept a book by Kate O'Brien – *Presentation Parlour*, a collection of reminiscences of O'Brien's early family life – close at hand as sustenance when, in her twenties, she had just moved with her husband, the writer Kevin Casey, out of their flat in town into a newly built suburban house in Dundrum, their Dublin home for the rest of their shared life.

Boland met O'Brien only once, when she was still in the early days of her writing life and O'Brien was in the final years of hers. She had come to their house for dinner. Casey was editing an anthology and after dinner O'Brien read aloud the excerpt from a novel she was writing called *Constancy* that he would include in his book. Boland recalled how O'Brien:

> [sat in their] new sparsely furnished house over at a sofa by the window in the front room. A frail woman by then, within a few years of her death, with a beautiful voice, and a deep theatrical cadence as she read out her fiction. The rainy October darkness unravelled from the half-built suburb where we now lived as I listened to her out towards the lights and boundaries

Eavan Boland contributed a personal assessment of Kate O'Brien for the Thomas Davis Lectures series marking the centenary of the Limerick writer in 1997.

of a city. This was the start of the 1970s. A city that was just on the edge of a new Europe, a new urbanity, a new identity – all attributes and adventures which were poignantly at odds with the grace and symmetry of Kate O'Brien's Ireland.

So much for the listener to take in, absorb and relish the detail and insight it offered about Kate O'Brien but also about Eavan Boland and how she saw the world and herself and Kate O'Brien in it. This is what a good radio talk does, and as time passes it becomes richer for the otherwise unreachable immediacy it offers of a time and present thinking that is past.

In 2020, during the Covid pandemic, RTÉ Radio rebroadcast and released a podcast of the 2002 Thomas Davis Lectures series led by Irish geographer Jim Hourihane entitled 'Engaging Spaces: People, Place and Space from an Irish Perspective'. The subject had present pertinence because personal movement was temporality limited. Contributors were scholars Drs Patrick Duffy, Ruth McManus, Desmond A. Gillmor, Mary Cawley, Des McCafferty, Yvonne Whelan and Joseph Brady. Between them their talks offered connectedness with space and place through relatable references to local house dances in 1930s Monaghan, observations of American writer Carson McCullers, who wrote that 'to know who you are, you have to have a place to come from', the first space tourist, Dennis Tito, who described his journey as 'a trip to paradise', and the sticking glue of the GAA club structure, which from early on followed the boundaries and borders of the parish and townlands to ensure local allegiance to the club jersey. Cultural and literary references through the talks included writers Lady Gregory, Wilfred Owen, John B. Keane, Karl Marx, Bill Bryson and John Denver. The lectures continue today as the Davis Now Lectures, providing the listeners with plenty to ponder.

Still and all, as historian and former editor of the *Irish Press* Tim Pat Coogan put it, writing in 1976 about his memories of the first fifty years of RTÉ Radio, 'the appreciative welcome the Thomas Davis Lectures received from listeners required to be related culturally to the even more widespread reception given to presenter Din Joe's invitation to *Take the Floor*' (*Take the Floor* was a hugely popular radio programme famously known for having guests dancing on the radio).

Bowman began working in Radio Éireann in the 1960s. He worked there. Writing about that time he noted how 'broadcasting was being transformed. It was a time of innovation and creativity, and risk taking. But initially, this was all happening on a greenfield site in Donnybrook where the new television studios were being built ... it took Radio Éireann some years to get its act together [to compete with the new arrival]. Not only were listeners choosing to watch television over radio but so were many of the staff moving over too.' By the late 1960s programmers started to develop new formats that were better suited to radio, exploiting its immediacy and intimacy, in which presenters encouraged listeners to get in touch by telephone or post. One of these, launched in November 1968, was *Day to Day*, presented by Freda McGough. It was an early variation of weekday morning programmes we know today, presenting 'people of interest, places of note and events to watch in a daily diary of the air'.

First recording of the presenter of *Take the Floor*, Din Joe, Denis Fitzgibbon, at Radio Éireann studios, Cork, taken on 9 September 1957.

Broadcaster Freda McGough and producer Brian Reynolds worked on *Day to Day* in 1968.

In its first few months *Day to Day* chose a series of short recorded talks – instead of studio interviews or reports – to draw attention for listeners to the country's changing education system. Free secondary education began in 1967, when only 36 per cent of Irish sixteen-year-olds were still enrolled in school. Thousands of students were now able to remain on. In 1969 the department of education published a free guide for parents called *All our Children / Ár nDaltaí Féin*, which covered infant education to university and technological colleges and was distributed through the schools. In an interview with *Day to Day*, minister for education Brian Lenihan said that, though more educational opportunities were available than before, it meant little if people didn't know about them. The booklet was the basis for a series of talks on the programme on subjects, which included the role of parents in decision-making, the new primary-school curriculum, the design of modern school buildings, the school health service and the education of children with physical, emotional or mental needs.

Producer and presenter John Quinn came into RTÉ through education programming. On Quinn's *Open Mind* programmes, which aired between 1989 and the early 2000s, he spoke with a variety of Irish and international thinkers on evolving ideas around learning and education. His work was deeply embedded in one of the original values of public-service broadcasting, that of knowledge-sharing. The annual Open Mind Lecture, which Quinn introduced, combined a keynote speech by a guest speaker on a subject of their choosing that they were passionate to share with the public, with a Q&A from a studio audience, some of which were included in the broadcast programme. A similar approach was taken with more recent Davis Now Lectures. The inclusion of the listener represented a shift towards a recognition that the attentive listener in the audience could also extend what we hear on air. A perhaps more inclusive, dialectic approach to learning from one another?

Michael Coady, who wove his hometown of Carrick-on-Suir, Co. Tipperary, into his writing, spoke about poetry and place in

2000. He declared that the brief he gave himself was 'to ground myself in the experience of a particular place and the mystery of time, focusing on the lower Suir Valley and my attempts to write out of it' and about his need for 'a couple of energising obsessions to push on' and how his might be 'time and place involving chance and destiny ... and the way lives fall out'. Coady turned to Patrick Kavanagh for a manifesto – Kavanagh's 'readiness to see grace in all things', and especially in the ordinary or everyday – and how the naming of places and things and people makes them more than transitory. Indeed, such a manifesto has more than a little about it of what radio can do too, populated as it is every day by places and people and things, somehow further validated by having been on the air.

Celebrated folk and Irish traditional singer Maighread Ní Dhomhnaill was among fifty people from various walks of life asked in 2000 by producer Marian Richardson to present a talk on a person of significance to them for the RTÉ Radio Millennium series commissioned by Lorelei Harris called *A Giant at My Shoulder*. She chose her father, Aodh Ó Domhnaill from Rann na Feirste in Donegal, a collector with the Irish Folklore Commission, a teacher and a singer. She recalled how, one day, when sitting outside her daughter's school waiting to collect her, she switched on the car radio to RnaG. She was only half paying attention to it when on came her father's voice. He had been dead twenty-two years. The sensation of hearing it then and there in a car in Galway was unreal – the power of accidental listening to bring us elsewhere. She was completely stilled, afraid to move, sucked in every syllable. And when her daughter arrived, she explained that the voice on the radio was that of her grandfather talking about songs he had heard in his native Donegal. She was overcome with a great longing for him but also a great pride in the person that he was. And she thought of the way her own life had developed and that of her sister Tríona and her brother Micheál and how so much of their involvement with music was rooted in this man. (Tríona and Micheál were members of the trailblazing 1970s Bothy Band.) She knew before she was ever asked that her father was the giant at her shoulder.

Listeners to that series heard contributions from chefs and religious to artists and writers. Twenty-five years later, a new

generation did the same. Writer Kerri Ní Dochartaigh's choice was Virginia Woolf. After years away from a practice of daily journaling, she said that Woolf's intoxicating diaries offered her a way back in. Through sitting and being with Woolf's diaries, Ní Dochartaigh realised, as she put it, '[That] for some people, maybe particularly for creative women ... the quiet act of placing one's own day-to-day life on the page is transformative, deeply healing work'. In their way each *Giant at My Shoulder* programme is illustrative of what Michael Coady mentioned in his Open Mind lecture about needing 'a couple of energising obsessions to push us on'. Doesn't everyone know the infectious pleasure of hearing someone talking expansively and generously about something, whether or not the subject in itself is of particular interest to the listener?

(Left)
Alice Maher, whose giant at her shoulder was Louise Bourgeois.

(Right)
Aodh Ó Dhomhnaill with Sorcha Chonail, at his work recording in Co. Donegal.

It seemed natural that short talks or writing features spoken by the people who wrote them would make their way on air when RTÉ's music and arts station lyric fm began in 1999. 'The Quiet Quarter' was broadcast daily on *Lyric Notes* for over a decade from 1999 to 2009, produced by Eoin Brady and presented by Máire Nic Gearailt. Nic Gearailt said they offered listeners 'a few minutes of quiet reflection in the morning, followed by music to suit the mood' with a guideline to contributors to offer their response to something rather than a report on something itself. All written to be spoken. Frank Corcoran described himself as a 'sound man' in his contribution, living in the German city of

Hamburg, a place steeped in music and musicians for centuries. As a composer himself living there for much of his working life, he muses on the fate of Johannes Brahams, who carried the city's rejection of him, not once but twice, with 'rage and sadness, that regret that his native city wouldn't properly honour him'. And how today Brahams is Hamburg's most celebrated composer.

Finally, in the 2000s producer Marian Richardson also offered listeners a temporary change of tone in the live daytime schedules when she commissioned Joseph O'Connor to be a regular diarist on Radio 1's *Drivetime* programme. His subjects would vary from a tribute to John McGahern to remembering the summer of 1979. A selection of these musings became a bestselling CD called *Drivetime Diaries*. Olivia O'Leary subsequently also contributed a popular topical radio column to the show, while Paul Durcan had his Radio Diary at an earlier time of day on *Today with Pat Kenny*. A collection of these was also published as a book. Comedian Anne Gildea began her short talk on Radio 1's *The Business* in 2026 by declaring that she was 'a bit old hat and ... came of age in the era of "dear sir or madam"... aware of when to sign off as "yours sincerely or faithfully", when knowing your "i"s before "e"s, caps, full stops and commas, etc were not just expected, but regarded as a sign of civility and respect towards whomsoever one *perchanced* to be communicating with ... see what I did there, ended on a preposition and it somehow don't seem to matter anymore dadeo!' And so she began, before going on to expand on the declining use of correct grammar and sentence structures and what it reveals about the state of the world. She suggested it was coming from the top down, from some of the most influential people, 'especially if you are in the upper echelons of wealth and power it seems'.

Her remarks brought two things to mind. The first was a once long-running RTÉ Radio programme called *Dear Sir or Madam*, which featured letters from listeners on all sorts of topics and even awarded a prize to the 'letter of the week'. The second was one of Joseph O'Connor's radio diaries, in which O'Connor recalled visiting a café with his wife where the waiter greeted them with a 'Hi guys.' O'Connor remarked, 'I don't even consider myself all that much of a guy, but the wife really isn't one at all.' He was bringing the listener's attention to what he called 'the latest

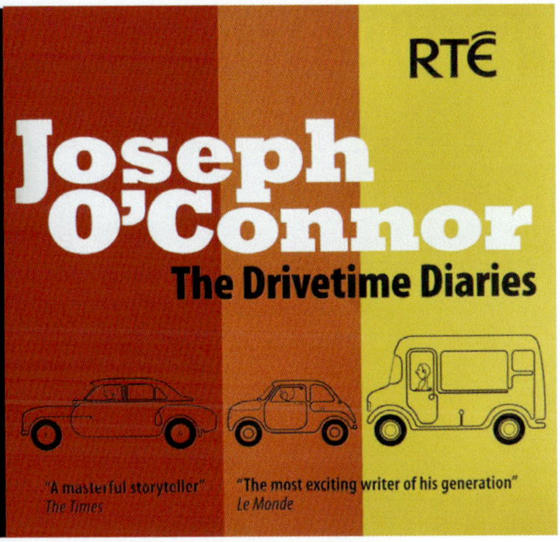

Joseph O'Connor recording *The Drivetime Diaries* CD in July 2010.

little virus to have weakened the language' along with 'Like do you know that kinda way?' and 'At the end of the day.' Gildea and O'Connor also reminded me of an essay called 'Continuity', which Ted Sheehy read on *Sunday Miscellany*. It related to his mother, Radio Éireann continuity announcer Una Sheehy. Why? Because in his talk Sheehy described how 'a voice or sound ... carried by a radio signal is not trapped in, or by the time or place it is heard'. Somehow all the radio talks that have been broadcast on RTÉ radio are still out there, drifting through space, each just a little ahead of the next, just almost close enough to be chatting to one another. They may also live on in similar ways in the memories of those who listened to them.

PRESIDENTIAL INAUGURATIONS

SPOTLIGHT

On 25 June 1938 Radio Éireann broadcast the inauguration of the first president of Ireland, as it has done for every president since. I can still remember, as a child, listening back in wonder to the presidential oaths being taken as part of an LP released in 1976 to celebrate fifty years of Irish radio. Whatever the reason, it caught our imagination. The ceremony, the seriousness, the simplicity of it.

Dr Douglas Hyde was the unanimous cross-political party choice to hold the role for the first time. He was sworn in *as Gaeilge*, setting a precedent for his successors. The unidentified Radio Éireann commentator on Hyde's inauguration didn't conceal his consciousness of the significance and newness of it all, assisting listeners to *see* as well as hear the occasion in St Patrick's Hall:

> Everybody is glancing anxiously towards the entrance door ... with everyone just as tense as I am on this glorious occasion ... and I've just been told that he is on his way now, so the hall is rising ... I feel such a tenseness here that it would almost be sinful of me to say anything to you just now ... Now here he comes, evidently the greatest of dignitaries. And you'll hear the fanfare accompanying him ...They're going slowly up the aisle now, up to the top end now and I'll tell you exactly when he comes to the steps ... he's mounting one, two, three. He's gone over to the right-hand side of the table. He's shaking hands now with An Taoiseach, Mr Éamon de Valera. He's moving around to the centre of the Hall. He's standing up in the very centre now.

That first inauguration was big news, and the coverage of it was an all-day affair. At 9 p.m. there was a short programme in Irish called *Ag Féachaint Siar* (Looking Back), in which Dr Eoin MacNeill spoke about Dr Douglas Hyde. It was followed at 9.10 p.m. by an hour-long *Commentary from the State Reception* in St Patrick's Hall, with Bernadette Plunkett and Brian Durnin. The late news of the evening of Douglas Hyde's inauguration included a recorded impression of the day's ceremonies. Newly acquired recording equipment in the GPO meant what was being broadcast could be recorded at the time and broadcast again later.

(Clockwise from bottom) Dr Douglas Hyde, first president of Ireland.

The RTÉ Archives acetate disc recording of the inauguration of the first president of Ireland on 25 June 1938.

Cover of the commercial LP record that includes the inaugural oaths of the presidents of Ireland up to 1976.

When Erskine Childers was inaugurated as the fourth president on 25 June 1973, RTÉ radio broadcast an ecumenical service from St Patrick's Cathedral before the ceremony held in St Patrick's Hall in Dublin Castle. RTÉ political correspondent Joe Fahy wrote in the *RTÉ Guide* that perhaps 'the pageantry associated with the inauguration may serve to mark a new beginning in the relationship between the Presidency and the people'. The perception that the presidency was remote from the people was one of the main issues in the election campaign.

The thousands who paid respect to President Childers laid out in St Patrick's Hall less than two years after his inauguration illustrated the connection the public felt with him and the loss of such a highly-regarded figure to the country. Taoiseach Liam Cosgrave declared 21 November 1974, the day of his State funeral, a national day of mourning. He remains the only president to die in office. When RTÉ Radio came on air that morning it broadcast a combination of solemn poetry and music and a reflective programme, *Ómós don Uachtarán* (In Honour of the President), presented by Ciarán Mac Mathúna, before the funeral broadcast.

The president's funeral did not deter further deadly violence. Already experienced in Dublin and Monaghan earlier in 1974 and long familiar in the North, it now extended to Britain. That night, as a radio programme entitled *In Memoriam* paid a final tribute to Erskine Childers, news was breaking that seventeen people had been killed and one hundred and twenty others injured in two bomb explosions in the centre of Birmingham. The atrocities of 21 November 1974 came to be known as the Birmingham Bombings.

Childers was the son of 1916 executed patriot Robert Erskine Childers and elected to the Dáil as a Fianna Fáil TD for Athlone–Longford in 1938. He had overseen the liberalisation of Radio Éireann in the 1950s with the establishment of Comhairle Raidió Éireann and its move towards independence from his own department of posts and telegraphs.

President Childers and his wife, Rita, during his inauguration as fourth president of Ireland on 25 June 1973. RTÉ provided live television and radio coverage of this inauguration.

(Opposite page) The public came out to pay respects to President Erskine Childers.

SPREADING THE WORD

CHAPTER 5

To paraphrase Louise Glück's idea that we see only once, in childhood, and the rest is memory: perhaps I truly *heard* only once, in my 1970s childhood, and have remembered it ever since. Repetition may have something to do with it too. I am thinking of the Angelus.

On Saturdays my siblings and I went over to Mamó, our grandmother. There, in her house, we witnessed her six o'clock right-foot toe-tapping genuflections to the *bong bong* of the Angelus coming from her transistor radio on the armrest of her chair. Restless to be away home at this stage in the day, we would have long since devoured the Rolos and red lemonade she had got in for us. Her right arm stretched out for her hand to hold the mantelpiece to steady herself. She was elsewhere for the run of the bells. Bent on prayer. She rose up and down again with each expression of 'Jesus' in her Hail Marys until the ritual ran its course. The *Sé do Bheatha, a Mhuire* (Hail Mary) we knew had '*Íosa*' in it. Her version in English detached us from what she was saying. The detail of her eyes being closed for the duration added to the drama. We in turn held on to the arms of the couch opposite. In unison, to put down time, we imitated her. We thought we were hilarious.

And here's the curious thing. There was no element of religiosity present in what was on air on the opening night of 2RN. Phrases such as, 'Let's pause for the Angelus' or 'But first, the Angelus' came later, quite a bit later. Not until 1950 in fact, during the year Pope Pius XII had designated the Holy Year when, on 15 August, the Feast of the Assumption, the bells rang for the Angelus on Radio Éireann for the first time. It was agreed between the radio station and the Archbishop of Dublin that the bells would be broadcast live at exactly six o'clock by an automated system connecting the GPO studios and the pro-cathedral across the way. They would be set to peal at a pace of 3–3–3–9, allowing for them to chime with those who wished to recite the related prayers.

They haven't stopped since. And every so often, they attract public comment. A self-declared atheist in *The Irish Times* on 8 September 2020 went on to say that:

> The Angelus, along with long spells of bell ringing or calls to prayer from a mosque, seem to me to not just be calls for reflection, but a strident demand for attention from the male hierarchies of the various religions. The playing of the Angelus here in Ireland is a daily reminder of the theocracy that ruled the State for so long. Maybe Alexa could be made to respond to 'Alexa, please play the Angelus' instead of so many of us having to shout 'Alexa, off!' between the main weather forecast and the news every evening?

A Radio Éireann Ford Anglia estate car with the St Brigid's logo on the side, photographed in the grounds of the RTÉ in Donnybrook, Dublin, in April 1966. The visibility of the station was important to let listeners know it was out and about.

Another letter in the same newspaper on 24 August 2022 commented on the broadcast being offered as a time of reflection for people of all faiths, and one suggested that, 'rather than continuing with the piercing and repetitive bong sound from the bells, would it not be better to air something a little more soothing? How about the tintinnabulation of wind chimes blowing in the breeze?'

A hundred years ago producers decided that the best course of action was sometimes not to broadcast at all. In 1928 a three-day closure of 2RN during Easter Holy Week was announced across the newspapers with the Gaiety and Queen's theatres in Dublin city centre, the Pavilion in Dún Laoghaire and others advertising that they would be closed too. The following year, 1929, readers of international publication *World Radio* of 23 March were told not to be surprised if on Good Friday they were not able to tune in to some familiar stations (showing how people did listen to many stations). Several would close for the entire day or at least limit transmission. Fond as the Irish are of superlatives, they would have liked to read that, 'the largest break will be made by Dublin which will close down from Wednesday until Saturday morning'. By the time I was a child, they seemed to broadcast solemn liturgical music for hours, before gradually secularising the offering – radio's singular capacity to establish the sound for a country's collective immersion into a given mood played out on air.

2RN, along with thousands of radio stations worldwide, carried the inaugural broadcast of Vatican Radio Station in 1931. For the first time in history the pope's voice was heard all at once across different time zones. This synchronised listening was reported in the press. For believers, it must have matched the awe of those who decades later heard and saw the first astronauts talk back to earth from the moon. Crowds gathered outside Brabants & Co. of Grand Parade in Cork City. The electrical supplier premises set up speakers from the upper windows, much as it did for match broadcasts during those years. A photograph catches the communal experience of radio-listening before personal sets became commonplace; the gathering echoes the habit men had

Crowds gather outside Brabants & Co. on the Grand Parade in Cork city to hear the voice of a pope for the first time on radio when on 7 February 1931, 2RN carried the relay from the opening of Vatican Radio.

of standing outside Mass, being there but not fully committing to it. Where were the women on this momentous occasion of gathering to experience the pope live? There in the middle, looking directly at the camera, a man nonchalantly smokes a cigarette, something he wouldn't feel free to do in the actual company of a religious figure in flow.

On 21 February 1931 the *Evening Standard* correspondent captures the hushed thrill of radio from almost a century ago, reporting that clergy, laity and journalists (and probably also the staff not on duty in the station's control room) gathered in silence in the waiting room in 2RN to listen together to the broadcast:

> All were waiting patiently for the voices that would issue from the 'loudspeaker' beside the wall. Atmospherics crackled, and queer noises ran through the air, suggesting the mysterious depths of space, the ocean of ether, that lay between Dublin

and the city of the Popes. A voice sounded – that of the Vatican City announcer – and spoke in several tongues. We distinguished Italian, French, German and English. It told how the Holy Father was proceeding to the broadcasting chamber. Again, stillness fell.

A second voice, beautifully clear, sounded. It was that of Italian Senatore Marconi, the inventor of wireless telegraphy. We were listening to the man whose brilliant inventive genius had given to humanity the most wonderful of all conquests of nature – the man to whom we owe it that thousands and thousands of lives have been saved at sea, and that the remote ends of the earth are knit in instant communication. Senatore Marconi spoke first in Italian, then in English. 'I have the pleasure and privilege to announce that the Supreme Pontiff, His Holiness Pope Pius XI, is about to broadcast from the Radio Station of the State of the Vatican City a message of peace and benediction that will be heard by all the world.

For nearly twenty centuries the Popes have sent their inspired messages to all peoples, but this is the first time that the voice of a Pope has been heard simultaneously in all parts of the globe ... The voice of the Sovereign Pontiff followed, and every listener in that chamber among the roofs of Dublin gripped by wonder and emotion. This was the Pope's own voice that sounded in our city! The same thought was thrilling companies of people in every corner of the land, from end to end of the inhabitable globe ... For fifteen minutes the Holy Father Spoke and at the end every one of us could recognise the words of benediction – In nomine Patris et Fili et Spiritus Santi – and knew that the Pope was signing the cross to all the world: we made the Holy Sign, and silence fell upon a room that was awed and humbly grateful.

In June 1932 the International Eucharistic Congress was held in Dublin in commemoration of the 1,500th anniversary of the mission of St Patrick to Ireland. By this time over 30,000 broadcasting licences had been issued in the Irish Free State.

The Eucharistic Congress brought the radio into many people's homes for the first time.

Many multiples of this number would hear the live broadcasts of the Congress. Athlone-born international star tenor Count John McCormack sang 'Panus Angelicus' live at the Congress's High Mass, which was attended by over a million people in the Phoenix Park. It was his second time in full flow on 2RN – the previous year he had sung at the first Mass to have been celebrated at the Rock of Cashel in three hundred years. That same year, 1931, he also starred as lead in the Hollywood film *Song of My Heart*. Years later Mamó Eile, my other grandmother, would recall to us children 'beautiful John McCormack's voice' coming from the speakers that had been distributed all over Dublin to carry those broadcasts. She was a woman who relished emotion in a song, the first person I knew to flick stations to find what she wanted to hear.

The State provided what was necessary to broadcast the week-long Congress in its entirety to the country and the world. Animosities between government and opposition, their deep divisions over how to bring about a united Ireland that caused such a bloody recent civil war, were put aside. Catholicism, and its capacity for ritual and ceremony, was recognised and seized.

The Congress began on 20 June with the opening reception of the Cardinal Legate, the highest-ranking office of the Church after the pope. This reception was held at a temporary building on Strand Road in Sandymount, at the recently extended city boundary of Dublin that now included Rathmines and Pembroke. The Cardinal Legate arrived in Ireland through Dún Laoghaire port a little further south of the city. The Congress ended the following Sunday with his final address after the celebration of Solemn Benediction on O'Connell Bridge in Dublin city centre.

Papal Legate Cardinal Lorenzo Lauri addressing the citizens of Dublin at the 31st International Eucharistic Congress in 1932. On his left are Lord Mayor of Dublin Alfie Byrne and Taoiseach Éamon de Valera, and to his right the Lord Mayor's Herald and other officials in ceremonial costume.

Marlborough Street, Dublin, 1932, decked out in bunting for the Eucharistic Congress.

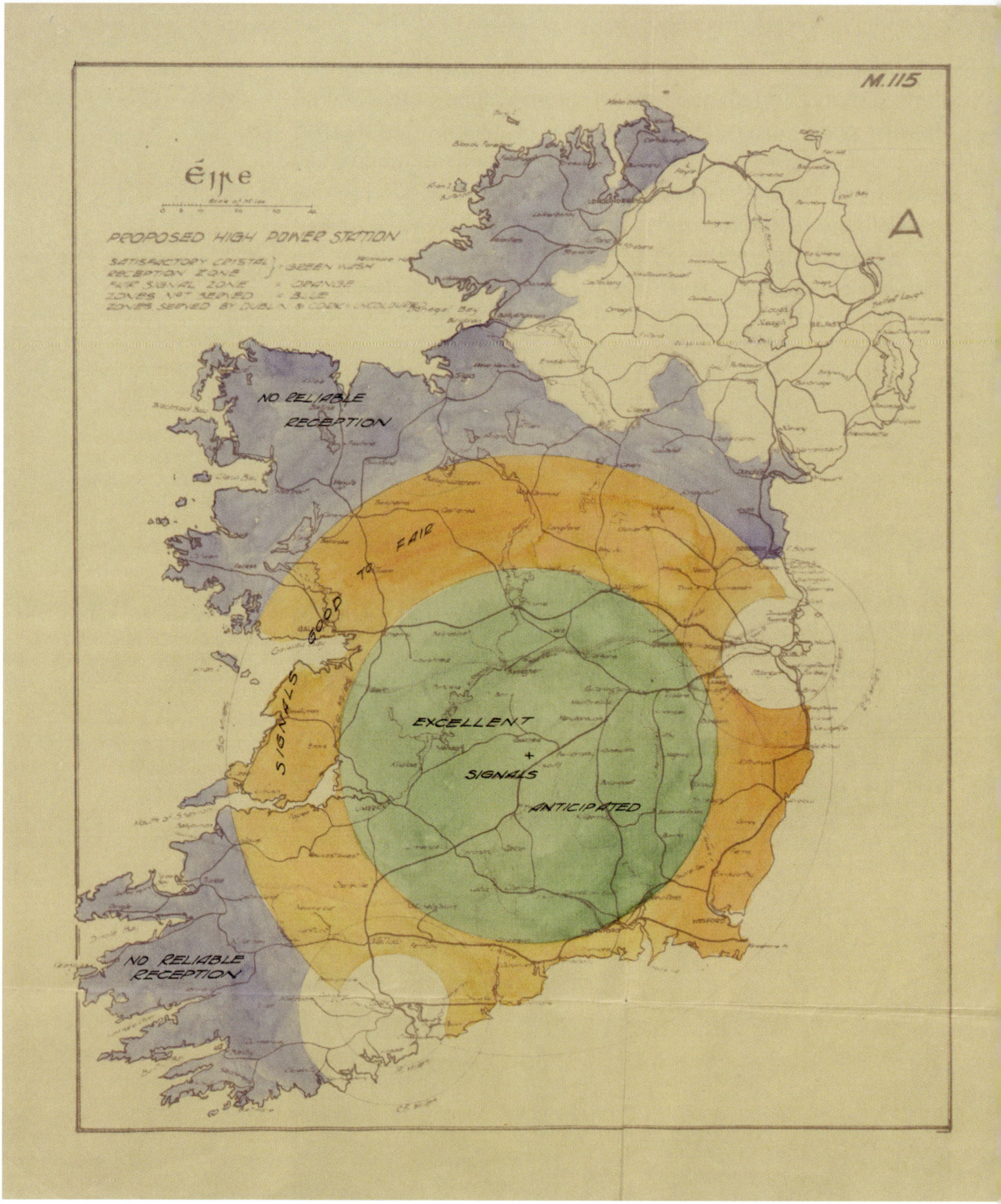

A 1926 map illustrated that Athlone was the ideal location for a national broadcasting transmitter, with a note advising that 'the areas of various grades of reception will not be perfect circles but will be somewhat irregular in shape. The average results should [be] close to those indicated.'

Additional broadcasts included receptions in the pro-cathedral and Dublin Castle, ceremonies and pontifical High Masses in the Phoenix Park, one with a choir of 2,700 children and another with a choir of 500 men and boys. Listeners also heard the voice of Pope Pius XI from Rome during the week. Ordinary evening programmes continued to be broadcast on the radio, along with announcements for Congress visitors in the mornings and additional talks on the proceedings of the Congress in the evenings.

Though still under construction, the powerful Athlone 60kw radio transmitter was temporarily brought to life, draped with tarpaulins to protect it from the elements – the clergy even appealed to people to pray for good weather – and shut again once the Congress was over.

The mast at Athlone, still under construction in 1932, was temporarily opened to broadcast the International Eucharistic Congress so that proceedings could reach as many listeners as possible in Ireland and internationally.

(Opposite page) Statue of Christ in the grounds of St Nicholas of Myra church on Francis Street, Dublin, in the 1950s, erected to commemorate the 31st International Eucharistic Congress.

In February 1939, following the announcement on the news at 1.30 p.m. of the death of the pope, Radio Éireann closed. It re-opened in the evening only to broadcast a tribute by Cardinal McRory of Armagh. A month later the Coronation of Pope Pius XII was relayed, courtesy of Vatican Radio, and broadcast live by Radio Éireann. The relationship between Irish radio and Vatican Radio had been cemented.

Before the Second World War, religious services had become a regular, if not yet continuous part of programming. During the war, sermons – like political speeches – were checked before broadcast to ensure nothing could be interpreted to show anything other than neutrality. On one occasion, just before going into the pulpit and during a live broadcast, a clergyman was given a note with the instruction by his bishop for it to be read. The note attacked Hitler on air. The minister for post and telegraphs, in a subsequent Dáil debate, stated how it 'might have got us into considerable trouble'. Station director Séamus Ó Braonáin used the excuse that he needed to see the sermons in advance of broadcast to ensure they kept within the time limit. The church authorities didn't like this being asked of them – probably knowing that the request was being made so that the scripts would be checked. The story goes that the Church showed their dislike of the request by having the next sermon overrun by some forty minutes. One can't help thinking that overruns by sports events could cause different tensions.

In 1948 Mass from different ecclesiastical colleges became a regular part of the schedule. Charles Kelly, by then station director, announced in the *Radio Éireann Yearbook* that this was 'to help people, who through age or infirmity are unable to be present at Mass, to participate spiritually in the offering of Holy Sacrifice'. His statement continued: 'The broadcast is deeply indebted to His Grace, the most Reverent Dr McQuaid, Archbishop of Dublin and Primate of Ireland, for the cordial co-operation which has made these weekly broadcasts possible.' Kelly wanted to make it clear that the station was providing something other than a religious service, and that it was for that

THE

reason that he was grateful for the cooperation he mentioned. When talks began about the Angelus being broadcast Kelly also objected to any suggestion of the Angelus prayer being recited on air.

Pope Pius XII declared 1950 a Holy Year. Uachtarán na hÉireann Seán T. O'Kelly led a national pilgrimage to Rome with the Primate of Ireland Cardinal D'Alton of Armagh. P. P. O'Rahilly of the Radio Éireann newsroom travelled to Rome with a small crew to bring first-hand accounts of events to listeners at home every evening. Indeed, on 29 April it was arranged that, through a link-up between Vatican Radio and Radio Éireann, Uachtarán O'Kelly reported his direct experience of his audience with Pope Pius XII to Ireland. Radio Éireann made history by sending a van equipped with tape-recording facilities to Rome. Necessity is the mother of invention and the team transported their equipment around the crowded streets of Rome in a home-made conveyance using the wheels of a pram – an innovation that showcases both the team's ingenuity and their underfunding. Later that year, Radio Éireann broadcast a series of programmes based on the material.

In the 1960s, Seán Mac Réamoinn, an expert in Liturgy, as well as on cultural matters in general, regularly reported from Rome on Radio Éireann about the Second Vatican Council, the gathering attempted by Pope John XXIII to modernise Catholicism. Many recall parents getting all het up about what they heard being reported from Rome, one turning off the radio and the other turning it on again. Mac Réamaoinn's own programme, *The Church in the World*, was listed in the *RTÉ Guide* as being broadcast from Rome on Sunday, 22 October 1965. The same evening, an tAthair Diarmuid Ó Laoghaire, SJ presented the programme *Aitheasc an*

Included in Radio Éireann's day sheet of St Patrick's Day 1962 is the radio address by President Éamon de Valera from Vatican Radio. He was in Rome for the final event of the Patrician Year.

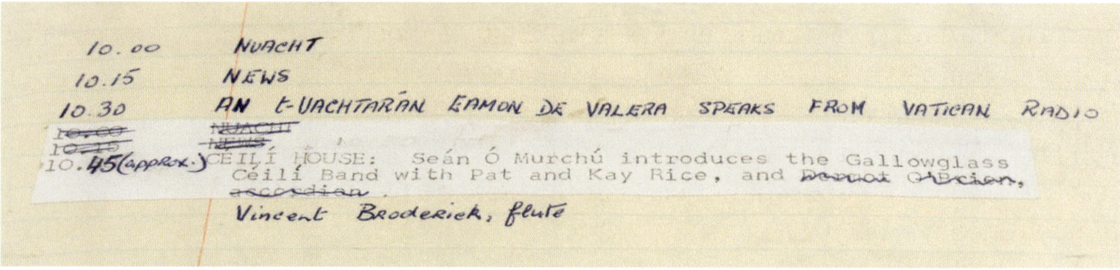

10.00 NUACHT
10.15 NEWS
10.30 AN t-UACHTARÁN EAMON DE VALERA SPEAKS FROM VATICAN RADIO
~~10.00~~ ~~NUACHT~~
~~10.15~~ ~~NEWS~~
10.45(approx.) CEILÍ HOUSE: Seán Ó Murchú introduces the Gallowglass Céilí Band with Pat and Kay Rice, and ~~Dermot O'Brien, accordian~~.
Vincent Broderick, flute

Domhnaigh (Sunday Address), a contemporary religious programme in Irish.

Anglican theologian Philippa Chavasse was the first woman to contribute to RTÉ's *Thought for the Day*, the weekday radio series consisting of brief early-morning talks by guest contributors. It ran for decades.

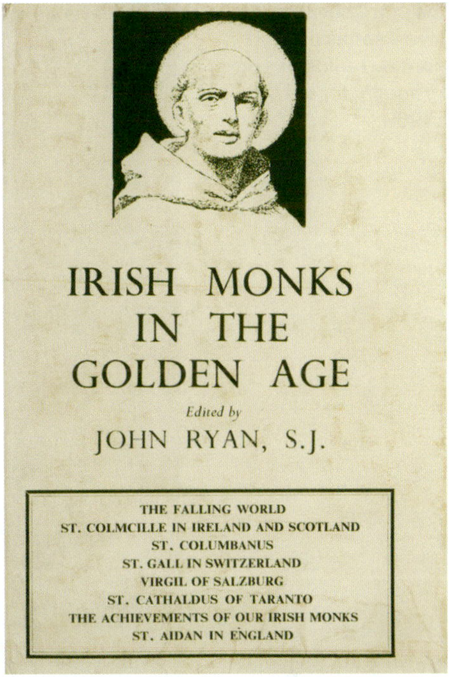

The Thomas Davis Lectures series *Irish Monks in the Golden Age* was led by Reverend John Ryan SJ, who edited the 1963 book of the talks.

Ahead of her broadcasts, Chavasse featured in the *RTÉ Guide* of 28 March 1969, where she was described as a woman of 'great charm, vivacious manner and a piquant sense of humour who unexpectedly turns out to be an authority on theology'. Why her authority would be unexpected is not clear and said more about the prevailing attitude that women could have an acquired knowledge, perhaps particularly in an area like theology, than anything about what listeners could expect to hear from her talks on air. English-born Chavasse was educated in a non-denominational school alongside Catholics, Hindus, Quakers, most Protestant denominations and many nationalities. She was an Anglican and felt the diverse community brought her an understanding of people, nations and religions other than her own. Expanded versions of her radio talks were published in *The Irish Messenger*. *The Irish Messenger* was a slim, red-paper-covered magazine I recall from my Mamó's house long ago, its cover often carrying a photograph or drawing of a saint-in-waiting such as Blessed Oliver Plunkett or a more decided image of the crucifixion at Easter.

Raidió na Gaeltachta went on air on Sunday 28 March, Easter Sunday 1972, as featured in Chapter 7. It followed a recorded message by President de Valera, who was now 89 and blind and not in a position to travel. De Valera's ambition – articulated from the 1940s and by others before that, though not delivered until decades later – was the provision of a radio station for the Gaeltachtaí and for Irish speakers. 'Aifreann Uí Riada' was sung by a Connemara choir, Cór Chúil Aoidha, and a choir from Tallaght, a choral embodiment of sorts of the station's motif, the three-swirled triskelion.

Philippa Chavasse gave a series of short morning radio talks on *A Thought for the Day* in 1969.

The blessing of the headquarters of RnaG in Casla, Co. Galway on Easter Sunday 1972.

The Mass was celebrated in Teach Phobail Mhic Dara, the parish church of An Cheathrú Rua, the church nearest to the new station's Casla headquarters, still under construction at the time. Somehow, it seemed natural that the station should be first celebrated in a building whose translation into English means 'the house of the community'. I sometimes think that composer, arranger and sometime RTÉ Radio broadcaster Seán Ó Riada, whose bust by the Cork sculptor Séamus Murphy graces the inner courtyard of the Radio Centre, keeps a cocked ear from there that RTÉ Radio captures what he called '*an rud sin ar a dtugaimíd dúchas*' (that thing which we call heritage).

The canonisation of Blessed Oliver Plunkett was broadcast from Rome on RTÉ Radio and Television on Sunday 12 October 1975.

The sixth-century journey of St Brendan the Navigator caught the imagination of explorer Tim Severin and became an RTÉ Radio programme in 1976.

The Salvation Army band, Dublin. *Morning Service* was broadcast from the Salvation Army Citadel in Rathmines on 29 June 1975.

While a Catholic service came to be broadcast on Radio 1 every Sunday morning, less prominent beliefs shared a weekly rotation for broadcasts. *Morning Service* was broadcast for the first time on 29 June 1975 from the Salvation Army Citadel in Rathmines, Dublin, for example.

To a Nunnery Gone was a documentary presented on 27 April 1979 by journalist and soon-to-be novelist Deirdre Purcell. The programme, described as a 'women's documentary', focused on the lives of nuns as well as their reception in the wider community. She visited numerous different orders to see at first-hand how they lived. Purcell, like many of her generation, was intrigued by nuns, having been a boarder at a convent school in Crossmolina, Co. Mayo. She went so far as to consider becoming a nun in her teens, though was glad she took the route she did in life, admitting to listeners that she thought her experience of them was not universal. Time would tell.

Deirdre Purcell in studio in the Radio Centre working on her documentary on the lives of nuns, first broadcast in 1979.

The *RTÉ Guide* cover of 29 September 1979 anticipated the arrival of a pope into Ireland for the first time.

The 1979 coverage of the first visit of a pope to Ireland was the biggest ecclesiastical broadcasting operation RTÉ had ever produced. For the 1932 Eucharistic Congress, the department of posts and telegraphs had only four telephone lines to carry signals from Dublin to the new transmitter in Athlone. This time, on-air radio alone required twenty-five reporters, as well as those reporting in Irish on RnaG, which offered its own full coverage. Between the public events, it broadcast a programme called *Fáiltimís Roimhe* (Let Us Welcome Him), offering listeners views of people from other churches in Ireland about the pope's visit.

Pope John Paul II being welcomed to Ireland at Dublin Airport by, from right of the pope: Maureen Lynch, wife of then Taoiseach Jack Lynch, An tUachtarán Patrick Hillery, Hillery's wife Maeve and Cardinal Tomás Ó Fiaich.

Communion servers during the Mass celebrated by Pope John Paul II in the Phoenix Park.

A child's return train ticket from Sandycove to Ashtown on 29 September 1979 for the Mass to be celebrated by Pope John Paul II in the Phoenix Park as part of his visit to Ireland.

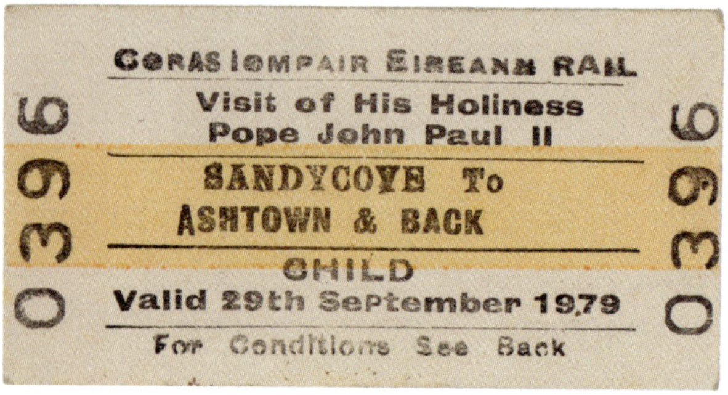

Pope Francis waves from his popemobile to crowds gathered in Phoenix Park.

Almost forty years later Pope Francis visited Ireland as part of the World Meeting of Families in August 2018. The pontiff had several engagements in Dublin, including meetings with Uachtarán Michael D. Higgins and Taoiseach Leo Varadkar. However, it was his private meeting with survivors of clerical sex abuse that most clearly illustrated that Ireland was a different country than that visited by his predecessor, Pope John Paul II. RTÉ Television, Radio and Online provided extensive live coverage, discussion and analysis.

In 1988, nearly ten years after Pope John Paul II's visit and ten years before the Good Friday Agreement, Dick Warner's series *Silent Minorites* began on Radio 1, featuring conversations with Protestants reflecting that the main barrier to a peaceful solution to what he called 'the National Question' was the view that 'Home Rule Is Rome Rule' and how discussions around the constitution relating to the right to life, abortion and divorce could reinforce that view. Many who spoke to Warner could not understand the

Joe Duffy and panel in the mobile studio in the Phoenix Park for radio coverage of the pope's Mass in 2018.

misconceptions many Catholics had of Protestants, or that the Queen of England was the head of their church when she was only the Head of the Anglican Church in England and Wales. The Church of Ireland, they said, was independent and self-governing and, like the Catholic Church, divided by a border.

Following Vatican II, religious books and their authors became valuable guides for listeners engaging with related discussions such as those featured in *The Church in Action* series presented by Andy O'Mahony. Wicklow-born Elizabeth Hamilton – author, biographer and novelist – was introduced as one of the more interesting new writers. Her autobiography *I Stay in the Church* told how she had turned from Protestantism to Catholicism at university and stayed with it despite some disillusions with parts of it. She also wrote about how reading theologians outside the Catholic Church – including Paul Tillich, Dietrich Bonhoeffer and Bishop John Robinson – influenced her to stay with Catholicism. The intention of *The Church in Action* was to offer listeners an insight into the spiritual life of the modern Christian. By 1999 Mary Finn in the *RTÉ Guide* of 25 June 1999 introduced the series *Soul Survival* to be 'for and about the care of the human soul'. The first guest was poet Brendan Kennelly. *The Godline*, presented by

13. 04. 20

Dear Director,

I want to say a sincere word of thanks to you and all your staff for the wonderful job you have done since the Country went into 'Lockdown' and people could no longer get to church. During Lent and especially at Easter it meant so much to be able to have daily Mass and to know that we were praying with others for our Country and world.

I had thought that perhaps the Service would finish at Easter but was so pleased this morning to discover that it was continuing. Thank you.

A sincere thank you to all the priests who are celebrating the Masses and to those from other faiths who are sharing words of Comfort and hope.

May God continue to bless all at R.TE. your families and all involved in the broadcasts. and keep you safe and well,

Letters of thanks in 2020 for broadcasting Mass during Covid.

Cover of the card of thanks.

13/7/20.

Dear Sir/Madam.

I wish to thank you for broadcasting Mass at 10.30AM each morning. It brought alot of ~~got~~ joy to the "oldies" like myself. Would you consider continuing the practice even when things go back to normal?! It would mean so much to so many people who cannot make it to Church. I am 92 myself and find it difficult to walk. Maybe a Saturday morning Mass could be added in!

Anyway, thank you so much for what you are doing. God bless you all. Sincerely.

Another letter of thanks in 2020 for broadcasting Mass during Covid.

Terri Garvey, was also a regular radio weekend programme in the 1990s, also looking broadly at religious matters, in much the same way as Áine Lawlor's more recent series, *Divine Sparks*.

Finally, from time to time related subjects show up in unexpected places. I leave you with this childhood memory recounted by Mike McCormack on Radio 1's on *The County Measure* in 2024:

> The village of Leckanvy is on the edge of Clew Bay right under the Holy Mountain, Croagh Patrick. And in my childhood we could walk out our back door and keep going in a straight line up the side of the reek as we called it until we reached the church on top over which we had that marvellous view of Clew Bay, and which as children we were always told that there was an island for every day of the year. And in those days the annual pilgrimage to the top of Croagh Patrick was done by night and we would stand at the gable of our house and look out over to the pilgrim path which ran up by the side of Campbell's pub in which during the darkness became a stream of light. Lamps, torches, flashlights making a whole ribbon of light up the side of the mountain. Sometime in the mid-seventies, this night pilgrimage came to an end. It was felt by the bishop or whoever, that there was too much devilment and fun being had on the pilgrim path in the hours of darkness and not enough penance. I laugh at that every time I hear it. Too much fun being had on the pilgrim path.

Croagh Patrick, Co. Mayo on Reek Sunday, 1975.

THE NEW CONFESSIONAL

We cannot know what future generations will make of the last half-century's sharing of the profoundly personal on air by people whose public lives are well known to us. It seems that in an increasingly secular society, the media – broadcast and more recently online – has become the country's new confessional. The appeal of such radio is undeniable.

Before the broadcast the presenter advises on the sensitivity of the subject and the circumstances of the recording or live interview. Many listeners cite these interviews as the most memorable and impactful radio experiences of their lives. Perhaps there is some comfort, or some thread of connection formed by hearing that other people's lives are as complex and fragile as our own. It can also be cathartic for the guest to bare their soul to a nation they have a real relationship with.

Sometimes such radio catches the intensity of life on the cusp of dying. When Marian Finucane met Nuala O'Faolain for lunch in 2008, O'Faolain told her friend she was dying. Marian later said that, unbelievable though it was, once Nuala told her about her cancer, they started talking about doing an interview. Marian had interviewed Nuala a few times since the publication of her memoir, *Are You Somebody?* She knew that 'being interviewed live provided [Nuala] with an important means of understanding herself. It was as if answering questions under pressure marshalled all her considerable gifts, allowing her to speak with her wonderful blend of candour, wit and passion.' And she did. They were recorded within days of the unedited broadcast of the conversation; the stop-starts told their own truth.

O'Faolain talked about getting her diagnosis in the emergency department of a New York hospital. She had built her life there and now that life – with her 'beloved room', her friends, writing and Ireland – was suddenly reduced to 'wretchedness, such sourness with life, and fear ...' She wondered 'how we all get away with our unthinkingness' and suggested a record 'that you might play at the end, I know you don't go playing records, particularly for dying people but still ...' The song was 'Thíos i Lár an Ghleanna' (Down in the Middle of the Glen), sung by Albert Fry:

'Is a Íosa i lár na bhflaithis, tabhair arais an oíche aréir' (And Jesus in heaven, bring back last night). Marian Finucane finished the interview by playing the song for O'Faolain.

Nuala O'Faolain with friends at the Merriman Summer School in County Clare in the 1970s.

(Left)
Nuala O'Faolain in 1998 at the time of the publication of her memoir *Are You Somebody?* with her cat in her Dublin garden.

(Right)
Cover of the first edition of *Are You Somebody?*

On 18 January 2015 Miriam O'Callaghan's guest on her Sunday morning show was Leo Varadkar. It was his thirty-sixth birthday. She had often interviewed him about policies and crises but said this was 'not one of those interviews'. During the live interview, he came out as gay to the Irish public – the first serving Irish government minister to do so. The upcoming marriage equality referendum, legislation on surrogacy and whether to change the ban on gay and bisexual men donating blood were coming down the line. As minister for health, Varadkar said he wanted to be honest with people about who he was, but also wanted to be 'an equal citizen in my own country, the country in which I am a member of government and now I am not'. Many parents of gay children called to thank him for speaking out publicly. Others contacted the programme to praise Varadkar for his openness, even if they disagreed with him politically. He hoped that he could be seen as the same person as he was before speaking out. Though Varadkar said he felt better for having done it, he wasn't sure if the reaction online would be so favourable and hoped the next generation wouldn't have to come out on air in the same way.

(Left)
Leo Varadkar with Miriam O'Callaghan outside the Radio Centre after his live interview on 18 January 2015.

(Right)
The celebratory atmosphere in Dublin Castle when on 22 May 2015, Ireland became the first country to vote yes to marriage equality.

In September 2025 Brendan O'Connor introduced a recording made earlier in the week of 'a grown-up conversation that dealt with difficult themes' with Brenda Fricker, adding playfully that 'we don't normally do house calls'. Brendan described the setting for the listener: Brenda in her bed, her beloved dog Juno beside her, surrounded by pictures and possessions of a life fully lived.

Their conversation marked the publication of her memoir, *She Died Young*, and touched on loneliness, happiness and joy, mental health, rape, marriage, miscarriages, suicide and death.

She told Brendan she was 'afraid to cross over thresholds ... of moving from one space to another' and that she was a lovely person when she wasn't sad. They talked about the brutality of her psychiatric treatment – being over-medicated and physically punished for reacting to its conditions – the 'general lack of interest in mental health' and her praise for Dr Anthony Clare, who she said saved her and countless others.

She said she was happy to die quietly that night but that with her luck, she would probably live to 110. In the quietness of the moment O'Connor asked her if she would keep going. She said she would. She died on 16 July 2026, aged 81.

Brenda Fricker and her dog Juno at home with Brendan O'Connor in September 2025 during and after their recording.

OUT AND ABOUT

CHAPTER 6

We tend to think of radio as coming to us from a studio, and it generally does. Across the way from where I am writing, Radio 1's *Today with David McCullagh* is broadcasting live from studio S7.1 here in Donnybrook, Dublin. However, broadcasts often come from other places. From reporting breaking news and nationally significant events to bringing artistic and cultural performances into homes across the country, programme-makers have combined evolving technology with ambition and imagination to create live and recorded outside broadcasts. They have brought audiences closer to Irish life, culture and history as it happens.

Soon after Benedict Kiely turned eighty in 1999 producer Seamus Hosey organised an outside broadcast crew to record him at home reading from *The Waves Behind Us: Further Memoirs*, his new book focusing on his early writing life. I had just started as a researcher with Hosey. Being there, assisting with the recording, was my first time to work on an outside broadcast. A modest but memorable affair, learning on the job.

Kiely was a journalist, critic, teacher and fiction writer, as well as a regular broadcaster. In his prophetic novel *Proxopera* (1977), set in contemporary Omagh, a man is forced by masked IRA terrorists to drive a lorry carrying a bomb into the town while they hold his family hostage out the road. Kiely dedicated the novel to 'the innocent dead'.

And to think that a fatal bomb actually exploded in Omagh on 15 August 1998, killing twenty-nine people and injuring hundreds, irrevocably changing lives and the sense of the place. It happened on Kiely's seventy-ninth birthday. The bombing truly mocked the recently signed and now historic Good Friday Agreement, an agreement that had ended such violence on the island up until then.

Working on this book has confirmed for me that the connection listeners have to a radio voice is, for the most part, personal, associated with times and emotions in their own lives, as much as about how and what the voice says on air. So it was for many listeners to mellifluous, Tyrone-accented Kiely's, said to be one of the most distinctive radio voices of Sunday mornings in the 1970s and 1980s. He described himself as 'a very casual labourer' of the station with 'accumulated memories of the place', one of many contributors who transitioned from recording in Radio Éireann to the new Radio Centre out in Donnybrook.

We recorded Kiely reading in his sitting-room armchair. As the work progressed, his partner and later wife, Frances, appeared every so often with a tray of tea and sandwiches, her contribution to pacing the work in hand to get the best out of him, before we set him up again to read another chapter into the microphone. This continued for two days. Those recordings extended Kiely's broadcasting years and listeners' experience of them into the twenty-first century. An aural record of a life now past.

A decade later I was producer of *Sunday Miscellany* and had introduced occasional outside broadcasts to the programme. After accepting an invitation from the 2010 Benedict Kiely Literary Weekend to record a live edition of the programme from the town's Strule Arts Centre, I introduced Kiely's voice from the programme's archive into the live recording in hand (he had died in 2007). The local audience welcomed their own with spontaneous applause. A familiar and loved voice brought full circle home, and out again into the world in a new radio programme. Something an outside broadcast could do through the technology and emotion of sound.

An early form of relay or outside broadcasting was part of 2RN's opening night in 1926. From the other side of the city at Beggars Bush Barracks a performance of the No. 1 Band of the Free State Army was relayed to the station by telephone lines and broadcast live. The effort paid off. *The Irish Times* of 2 January 1926, in its coverage of the event, drew attention to the organisational side of the relay and of preparation meetings that had taken place between those involved in the station control room, the Beggars Bush contribution of the band and the official

The Army No.1 Band during the RTÉ100 outside broadcast from the GPO in 2026. The same band performed on the opening night in 1926.

in charge of the microphone there. And though the fifteen-minute performance ran over time, and threw the programme off schedule, the newspaper reported that 'the band-playing was amongst the most enjoyable of the night's entertainment'. The report continued, 'the performance of the Army Band was a revelation as to how good wireless can be. The placing and balance of the instruments was better than has been done hitherto in Europe. This most difficult problem seems to have been solved at first go by the Dublin Station ... Even to a technician without a musical soul the Army Band was a real treat.' The station, it concluded, 'can claim an approach to perfection. May the attainment of it be their ideal, and may they achieve it!'

The station's first outside broadcast from Cork took place on Sunday, 30 May 1926. Like all outside broadcasts at the time, it was relayed by telephone lines to the Dublin station from where it was broadcast. The event comprised an entire evening of performances by Sam Joy, a well-known baritone singer, and the Butter Exchange Band, which took place in the Cork School of Music. Later that year, the Crusaders oratorio was relayed live from Sligo and broadcast by 2RN; however, the telephone line

A notice in the *Evening Herald* of 10 September 1930 announcing the live broadcast on 2RN the following day of the movie *The Love Parade* from the Capitol cinema in Dublin.

relaying the sound was patchy, and the Dublin station orchestra had to take over for a while before the station engineers cleared the line and the Sligo outside broadcast could be picked up again.

Many of the country's telephone lines, though good enough for speech, were inadequate for music or singing – although listeners were forgiving for the ambition of what was being attempted by the station. When the Dublin station relayed the *Messiah* from Waterford in December 1926, W. D. Hogan of the Radio House in Henry Street, Dublin, put up loudspeakers on the premises, and the sound was cheered by hundreds gathered on the street for the occasion.

By mid-1927 there had been twenty-four relays broadcast by the Dublin station from Waterford, Cork, Newry, Arklow, Sligo, Thurles, Dundalk and Maynooth, as well as from various locations in and around Dublin. What listeners heard from them included movie screenings, art exhibition openings (we can presume a speech or single voice description rather than the chatter of those in attendance), musical events, presentations of plays and concerts, as well as live sports coverage.

There were also live relays from BBC stations. The Dublin station learnt to be vigilant about what might be broadcast from there that could ruffle Irish listeners. One London relay was cut short when a sketch called 'Loyalty' was substituted for the one that was expected; in another, from Belfast, there was a similar cut when an old drinking song proposed a toast to the King. Short but successful relays from places including Berlin, Paris and Hilversum, as well as other short-wave stations in America and the Netherlands, were also broadcast.

There was a tangible sense that a special event had taken place in a report in *The Irish Times* of 20 November 1928 of the previous night's performance of Bizet's *Carmen* at the Gaiety in Dublin by the Royal Carla Rosa Opera Company. The report noted that, apart from the large audience in the theatre, many thousands of people enjoyed that night's production, which was broadcast in its entirety by the Dublin station. It was the most ambitious feat yet attempted by 2RN, and those responsible, the report said, were to be congratulated on its success.

By 1928 there were 26,406 licensed radio sets in the country, so the reporter may have been right about the numbers who heard the opera broadcast. It is not clear whether the writer was in the theatre or listened in elsewhere, but from the absence of a description of what the set and costumes looked like and the mention of listening on headphones, it may be that the writer heard the opera on a radio set.

Though the daily schedule was still only a few hours, experiments were ongoing with what listeners heard relayed. *The Irish Times* of 3 September 1935 featured a photograph from Leinster Branch's swimming gala held in Blackrock seawater baths, Co. Dublin, a glamorous swimming and water sports venue at the time. The photograph shown overleaf is of a Mr K. McCullagh, who 'broadcast' from underwater with a microphone. It has not been verified if this broadcast was locally done on a megaphone or broadcast for radio. Either way it illustrates the fascination with catching the on-location moment and sharing it with others. Underwater sound was only captured for the first time in Bermuda in 1932. So it was definitely still 'Something New', as the caption to the photograph said.

Even if the broadcast only lasted a few minutes, imagine the impression it must have made on listeners. It's hard to imagine now, but many Irish people only visited the sea once a year, if at all.

"CARMEN" BROADCAST.

EXCELLENT PERFORMANCE AT GAIETY THEATRE.

The Royal Carl Rosa Opera Company entered on the second week of their season at the Gaiety Theatre, Dublin, last night, when Bizet's "Carmen" again was produced with complete success.

Miss Rispah Goodacre had already created a very favourable impression by her singing as Fricka in the "Valkyrie," and her Carmen was eagerly awaited. She gave a very clever study of the gipsy, conveying rather fully the sensuous, yet volatile, nature of the character. Her singing was rich and full, and showed admirable technique. She should develop to great things in the future.

Mr. Ben Williams as Don José sang and acted the part to perfection. His voice was always under wonderful control and his singing full of expression and charm. The Escamillo was sung by Mr. Leslie Jones, and he carried the part with the touch of distinction which he always gives to his work.

Miss Pauline Bindley was the Micaela, and she made the character something more than that of the simple country girl without losing any of its charm. Micaela's song was daintily sung, as was also the little duet with the tenor.

Mr. Aylmer Buesst was the conductor of the night, and showed a thorough grasp of the work.

Apart from the large audience in the theatre many thousands of people enjoyed last night's production which was broadcast in its entirety by the Dublin station. Microphones were fixed in the wings of the stage, with expert wireless electricians in attendance, and the performance "came through" on headphones and loud-speakers with wonderful clearness. It was the most ambitious feat yet attempted by 2RN, and those responsible are to be congratulated upon its success.

A report in *The Irish Times* of 20 November 1928 on the relayed broadcast by 2RN of a performance of Bizet's opera *Carmen* by the Royal Carl Rosa Opera Company from the Gaiety Theatre in Dublin.

(I clearly remember, in the late 1970s, my mother's non-swimming Limerick friends visiting us in Dublin and squealing with delight at the novelty of being up to their knees in the sea.)

The *Irish Independent* of 8 July 1935 reported on a relayed broadcast of an opera in Irish performed by the pupils of St Louis Convent School in Monaghan town the previous night. *An Cailín Bán*, adapted from Julius Benedict's *The Lily of Killarney*, was 'translated into Irish and produced by the teaching staff of the Convent'. The report congratulated the pupils and their staff for the ambition of their achievement, that the broadcast benefited from 'a good telephone line from Monaghan and this ensured that reception would be perfect'. It continued that 'it was quite possible for listeners to follow the opera, even without a knowledge of Irish, or with only a slight knowledge'. The report concluded that, 'from the musical point of view, both vocal and instrumental, the production was all that could be desired'.

So how did telephone lines transmit sound from the location of an event to the station engineer's control room in the GPO? Radio Éireann's P. P. O'Reilly's article in the *Irish Press* of 16 September 1946 described the technicalities of setting up a live relayed broadcast, using the Requiem Mass for the late Archbishop of Cashel as an example. Radio Éireann authorities would inform the telephone section of their own department of post and telegraphs that it was proposing to broadcast. The telephone section would decide on the most suitable lines to use. Though direct ones would deliver the best quality, it was important not to block telephone lines for more than an hour and run the risk of annoying ordinary telephone-users. The lines used might take a longer, more circuitous route to Dublin. For the broadcast from Cashel, the outside broadcast team liaised with Dr Morris of St Patrick's College, Thurles (who also provided the broadcast commentary during the Mass) on where

SOMETHING NEW

Mr. K. McCullagh (Clonard and Clontarf S.Cs.) about to submerge with microphone to give one of the underwater broadcasts, which were a popular feature of the Leinster Branch's Swimming Gala, at Blackrock Baths on Saturday last.

Underwater outside broadcast.

best to place the microphones in the cathedral. The engineers connected them to an amplifier, which had control over their volumes and could be faded up or down as required. And that sound passed through the cathedral telephone line to Thurles and on to the GPO using the agreed telephone lines.

Outside broadcasts or relays described up to now have all been single or once-off broadcasts. *Question Time*, on the other hand, was a weekly Sunday night general-knowledge quiz launched in 1938 by Radio Éireann's John MacDonagh. MacDonagh was the station's first director of productions (and a brother of Thomas MacDonagh of Cloughjordan, Co. Tipperary, a signatory to the 1916 Proclamation). He had founded one of the earliest Irish variety touring companies and, as well as performing in Ireland, had toured the company in the US and Canada. He had written songs and scripts for the old fit-up circuit (in which theatre and entertainment shows toured to villages and small towns, performing in local halls or their own pitched tent) and understood the power and effect of bringing a show directly to people and getting them involved. Writing in the *RTV Guide* of 13 July 1962, Joan D'Alton, who worked with MacDonagh on *Question Time* before producing the programme after him, recalled how in:

School Girls in Opera—A scene from the opera, "An Cailín Bán," broadcast by the pupils of St. Louis Convent, Monaghan, from Dublin, Athlone and Cork last night.

A photograph in the *Irish Independent* 1935 of a school production of *An Cailín Bán*, described as having been 'broadcast from Dublin, Athlone and Cork' the previous night.

> the pre-television years listening to Radio Éireann's *Question Time* on Sunday nights was almost a national pastime. Whole families settled round the fire, picked out their favourite competitors – on an arbitrary fancy for their occupations, their voices, or their ages – laid innocent bets on their progress and

rooted solidly for them right through to their exiting final. Our family was no exception to the national rule, and in my gymslip-and-school-tie days, the highlight of Sundays was that lively hour en famille.

When rural homes were first electrified, one of the most popular appliances was a radio set as it was cheap and something which many people were already using in some way before.

In 1939, a year after *Question Time* started, Radio Éireann undertook the first listener research. It was the most listened-to programme by a fair margin. Irish-speaking Gaeltacht areas answered the same as the rest of the country, and this prompted the station to develop versions in Irish called *Tráth na gCeisteanna* and *Ceist agam Ort*, two of many Irish-language variations of programmes. The fact that *Question Time* broadcast live from venues around the country doubtless played a part in its huge popularity. Radio Éireann was learning the importance of being on the ground with listeners.

An attractive incentive for locations to host *Question Time* was the ticketed post-broadcast concert, with what was raised given to a local charity. Even the Irish Tourist Association sponsored the hosting of programmes at various seaside towns during the summer months.

MacDonagh understood the importance of having the right compere to connect with audiences and tried several before

Joe Linnane became the regular host. Joan D'Alton wrote how Linnane had 'crispness, charm, a ready wit, and a friendly way with him. Also – and this was important in those days, when each competitor had to sing a few bars of a song – he could play the piano. He was a seasoned actor – how many remember him in the early [Lennox] Robinson and [Sean] O'Casey plays at the Abbey? – and had made a name in radio with sophisticated songs at the piano.'

Though she had since left the station where she had overseen women and children's programming, Máiréad Ní Ghráda – an experienced writer, translator and editor – regularly provided the questions and answers for *Question Time.* The programme came from wherever in the country there were enough telephone lines to relay the programme to the GPO and where the area would not be cut off from emergency services – such as a hospital or a priest – for the duration of the broadcast. The many broadcast locations illustrated that the telephone system, like rural electrification, was slowly extending, changing the nature of radio programmes and reaching new listeners. Those listeners were themselves moving on from crystal or valve sets to plug-in electric sets. The first Rural Electrification Scheme ran between 1946 and 1965, when over 300,000 homes were connected.

Picture this: *Question Time* presenter Joe Linnane plunging through snowdrifts to the nearest functioning telephone exchange for an emergency hook-up. This was a regular occurrence in the cold winter months, when the reliability of telephone lines to relay outside broadcasts was challenged by weather. From the 1960s the laying of co-axial cables and copper-covered wiring facilitated much more reliable outside broadcasts.

Question Time was also a revelation of what could happen in a situation with a live microphone without a previously agreed script (being a quiz programme, it operated outside the censorship imposed on scripts during of the war years). In a broadcast from Belfast during the war, Joe Linnane asked a competitor: 'Who was the world's best-known teller of fairy tales?' Instead of Hans Christian Andersen, the competitor answered, 'Winston Churchill.' The audience broke into laughter, which was transmitted live on air. (There were questions about what had happened in the House of Commons at Westminster, and it was a long time before a similar live programme was broadcast again.)

As a measure of the success of a programme, Radio Éireann introduced variations on *Question Time* such as *Family Question Time* as well as a *Dublin-Cardiff Question Time* and *The New Question Time* that came from the GPO studios with well-known contemporary personalities Din Joe, Al Thomas and Jack Cruise.

When Radio Éireann acquired a mobile recording unit (MRU) in 1947 it was a game changer for the broadcaster. The unit enabled, for the first time, for quality recordings of traditional music and song to be made on location – where those who played and sang and introduced what they performed were at home – giving them their wonderful natural atmosphere. MRU staff gathered and shaped material for inclusion on cultural and societal programmes that captured a changing Ireland – a record of voices spoken and music performed and built upon ever since, generation by generation. These recordings enable us to listen to first-hand voices recorded three-quarters of a century ago. It is transformative. A passing on. A making new again. Not to mention the pleasure of those who heard the recordings the first time around.

Before the MRU, performers of traditional music and songs hoping to be on the radio were auditioned by the Radio Éireann director of music or someone who deputised for them in the station in much the same way as classical musicians were auditioned. This meant only those who could or were willing to travel to Dublin – if they were not already living in or near the city – had a chance to be even considered.

The uilleann piper Séamus Ennis was one of the first officers recruited to the MRU (as it happened, his father James Ennis performed on the uilleann pipes on 2RN on the opening night of the station in 1926). Séamus Ennis had already worked as a collector for the Irish Folklore Commission and knew the country intimately. He had cycled the roads during the war doing his work and knew where to find people to record – or people to suggest who he should record. This networking and local knowledge is still as critical today to make the best outside broadcasts.

Fred Desmond (Fricker) with the mobile recording unit van at the World Ploughing Championships in Killarney in 1952.

Seán Mac Réamoinn, also an early outside broadcast officer, described the first MRU vans as being as cumbersome as the county ambulances of the time, although officers became adept at managing the recording equipment. Mac Réamoinn explained that, although their work resembled the 'collect the fragments as they perish' motto of the Irish Folklore Commission, as he put it, 'they were programme-makers after all, not folklorists'. In 2001 former RTÉ chief librarian Diarmuid Breathnach said that, 'as far as music was concerned, for the first time there was the beginning of an archive, and all that material became very valuable in later years'. Indeed, it did, as the recording of performances continues today. The mobile studio of Radió na Gaeltachta and Radio 1 feature prominently during the annual festivals of the Willie Clancy Summer School in Milltown Malbay or at Fleadh Cheoil na hÉireann and Oireachtas na Gaeilge, wherever they are happening in any given year.

Soon after he started working in Radio Éireann's MRU in 1954, Ciarán Mac Mathúna made his first field trip. He gathered versions of tunes from different parts of the country, which came

(Left)
Radio Éireann engineer Ned Nugent with Máire Nic Eachmharcaigh (school teacher and singer) during an outside broadcast in Crolly, Co. Donegal in April 1955.

(Right)
From left: Radio Éireann's Seán Ó Tuairisg, Seán Ó hEochaidh, of the Folklore Commission and possibly Ned Nugent, Radio Éireann engineer in front of the mobile recording unit van in a fishing port in Mín Lárach in the Donegal Gaeltacht in 1955.

to be of great interest to musicians, scholars and followers of traditional music, as well as the public at large. His timing was fortuitous. Radios were becoming cheaper and more widely owned. Skipping a generation, younger people were beginning to take an interest in the music cherished by their grandparents, a general post-World War II phenomenon not limited to Ireland.

Here, radio played a crucial role in learning new tunes. Musicians recall sitting ready with their instruments, listening intently to broadcasts and often picking up a melody by the time it had been played through once. They knew that fellow musicians elsewhere would be doing the same, all hoping the tune would be aired again in the following week or two so they could refine it, fill in any gaps and truly make it their own. Commercial recordings, records and cassette tapes of versions and variations, or new material that often came under the influence of what was heard on the radio, would come a decade or two later.

Mac Mathúna had a lovely way of introducing musicians, giving context to their tunes and songs. There is laughter and joviality in his recordings; the resulting radio programmes must surely have been great company to listeners. Flute player Michael McNamara from Carrickavoher in County Leitrim bought his first radio in the early 1950s. He was mad for traditional music and looked after the batteries with great care before the family acquired an electric plug-in radio once they had electricity.

Phyllis Power, Gloria Greene, Seamus Kavanagh and Jack Cruise check the script for a Radio Éireann outside broadcast of *Holiday Hayride* at Butlins Holiday Camp on June 1953.

(His talk of charging the batteries would put you in mind of contemporary talk about charging the batteries of electric cars, how sometimes the charging would hold for longer than other times, so that you'd have to be clever with how you would use the radio, not listening to everything that was broadcast.) The McNamara family invested in a second battery:

> For the big occasion such as *Céilí House* of a Saturday night with Seán Ó Murchú, or *Ceolta Tíre* (Job of Journey Work) with Ciarán Mac Mathúna. They were programmes we could not miss. We came in out of the field from working, tailoring our day to suit the programmes. My father and myself took our tea at five o'clock. We went out to the field. He had a watch and he says 'in about ten minutes time Ciarán will be on'. We walked back. We listened to his programme and went out to the field and worked until nine.

When I first began listening to Ciarán Mac Mathúna presenting his programme *Mo Cheol Thú* on Sunday mornings from the 1970s (and on until the early 2000s), I didn't know the long legacy he was carrying, the journey work he had already put in collecting music and songs – from his beloved County Clare or his native Limerick and so many other places around the island, or indeed further afield from the Irish emigrant population in New York or

Britain. He would quietly, intimately relate how he had recorded the particular tune we had just heard in such and such a place in such and such a year, before going on to introduce a recitation by the likes of Blánaid Irvine or Neasa Ní Annracháin, of a poem or other that fitted the particular mood of the programme that morning.

A recorded outside broadcast was key to *Céilí House*, which began in the 1950s. Other similar and hugely popular programmes that were also typically outside broadcasts and came from local halls or houses included *A Ballad Maker's Saturday Night*, *The Rambling House* and *Take the Floor*. They came and went, but *Céilí House* still thrives today. It is difficult to imagine that in the 1960s the show could regularly gather 800,000 listeners. The story goes that in a local pub, when *The Late Late Show* was still being broadcast on a Saturday night, once the time reached half past ten, the television would be switched off for *Céilí House*. Here was music to make you move, to get up and dance to, in the kitchen or wherever you were. Here was the recording of an event in much the same way as today the programme website says it 'travels the length and breadth of Ireland and beyond, in search of a good session to bring each week to the many loyal listeners'.

For Séan Ó Murchú, *Céilí House*'s first presenter, being there, wherever a programme was broadcast live, was critical. He would share information with the listeners, peppering his introductions with anecdotes and histories of the places. National broadcasting burrowing deep into the local, giving it a universal significance in the process.

Breandán Ó Dúill followed as presenter, then Kieran Hanrahan and Tristan Rosenstock. Though *Céilí House* presenters and contributors have changed over the years, the convention of how each programme opens remains the same. Brief words of introduction by the presenter. A declarative whooping and clapping by those gathered, a demonstrative sort of 'We're here, we're on air, take it away!' Back to the presenter, who names the first tunes. In comes the playing. And the programme is up and running for near on an hour. The naming of local places and people and their musical instruments continues: everyone given their due moment, even when the first band might have seventeen players – the democratisation of radio in action. There is a palpable sense that one week you could be a listener, the next week a contributor. A simple, adaptable form for more than half the time RTÉ Radio has been on air.

In an archive interview with Pádraig Ó Méalóid in 1986, Mac Réamoinn said that the MRU was a shorthand for the invaluable work of the station of collecting music and song but that, just as crucially, going out with the microphones illustrated to people that their lives and experiences were significant. They had value. They mattered. Talking into a microphone encouraged people to take the first tentative steps towards self-expression. And although this seems like a natural thing to do today, in mid-twentieth-century Ireland it would have felt radical, paving the way for so much groundbreaking broadcasting of more recent decades.

Provincial News Round-up and *City News Reel* both started in 1957, one to reflect topical happenings around the country, the other for events in Dublin and its environs. The programme titles confirm the prevailing notion that everything outside the city of Dublin was provincial – curious, since the station's staff were from all over the country. Reports from correspondents were collected on new portable recorders, edited and introduced into the programmes for weekend broadcasting.

In 1959 presenter Karl Jones handed listeners over to P. P. O'Reilly, who was reporting on floods in the midlands. It is poignant to hear him describe where he was, 'on a junction of the Clonown Road just on the borders of Athlone near the Shamrock Lodge Hotel and the new chocolate factory. I can see the spire of St Mary's over there on my left and the twin aerial markers of the Athlone Transmitter ...' And there it was, mentioned in his report without undue consideration: the Athlone Transmitter, the central structure of Irish radio broadcasting at the time. And then back inside his car for shelter, he focused his microphone on the roof of the vehicle, letting listeners hear for themselves the pouring rain outside.

Nearly a decade later *Provincial News Round-up*'s presenter was Kevin McDonald. In the *RTÉ Guide* of 22 October 1965 he explained that the programme, in its entirety, was now being frequently broadcast live by the outside broadcast unit from different locations around the country, rather than having reports from around the country into the programme broadcast from Dublin. Many listeners, including Irish people in the north of England and in Wales, had been in touch to say they enjoyed the change.

The 1968 *RDS Annual Proceedings* reported on an innovation of its Annual Dublin Horse Show: 'the Radio Centre, organised by Radio Telefis Éireann. A central site was made available and a studio set-up from which live programmes were broadcast daily direct from the showgrounds.' New technology was changing how listeners could hear radio, so one of RTÉ's motivations in attending the Horse Show was to promote the new VHF (Very High Frequency) service to the 120,000 visitors who attended the show during the week. Not only did VHF promise a much higher quality than medium wavelength radio, but it also prevented interference to its transmission from other electrical equipment and other radio stations. RTÉ had a stand displaying VHF radios, a fully functioning radio studio and an information desk where visitors could also leave requests. Live music programmes broadcast during the show included *Rogha na mBan* (Women's Choice) and *Hospital Requests*.

Different comperes – Terry Wogan, Lorna Madigan, Maurice O'Doherty, Brendan Balfe, Mike Murphy, Joe Linnane, Bunny Carr and Una Sheehy – hosted the shows, while popular singers who dropped by and featured on air included Dickie Rock, Danny Doyle, Seán McDhonncha and Johnny McEvoy. RTÉ outside broadcasts from the Dublin Horse Show continued for many years, with similar arrangements put in place for the Spring Show and the annual National Ploughing Championships, as well as for Bord Bia Bloom, which began in the 2000s.

(Opposite page) The *RTÉ Guide* cover of 2 August 1968 promoted the first days of broadcasting by RTÉ Radio from the RDS Horse Show.

RTE Guide

PROGRAMMES AUGUST 3–AUGUST 9 Vol. 5, No. 32 :: IRIS RADIO TELEFIS EIREANN, AUGUST 2, 1968. PRICE 6d.

NOEL ANDREWS

BRENDAN BALFE

BUNNY CARR

VAL JOYCE

BRIGID KILFEATHER

RTE AT THE RDS

SEE PAGE THREE

LARRY GOGAN

JOE LINNANE

LORNA MADIGAN

MIKE MURPHY

MAURICE O'DOHERTY

JOHN SKEHAN

TERRY WOGAN

INSIDE: SPECIAL 8-PAGE PULL-OUT BALLAD BOOKLET

The new headquarters of the Irish Countrywomen's Association opened in 1968 at An Grianán in Termonfeckin, Co. Louth, from where Des Kenny hosted two episodes of *The Irish Gardener* in 1970. The programme hoped to encourage horticulture as a career for women and to highlight available ICA courses, as well as scholarships from County Committees of Agriculture and the department of agriculture at a time of growth and development in secondary and vocational education. This intention chimed with the station's public-service remit of providing information to listeners while entertaining them with a quiz in the programme. John Fanning, who accompanied the programme to An Grianán to report for the *RTÉ Guide*, described the progressive atmosphere of the place, including 'a young student wearing faded jeans and runners' who said she had 'no desire to marry "a glass-house man"' and intended to follow her own horticultural path.

Recording an outside broadcast edition of *The Irish Gardener* in An Grianán, the ICA headquarters in Co. Meath and broadcast in 1970.

By the 1970s RTÉ was well aware of the limitation of one station providing everything that could be expected of Irish radio. Meanwhile, the public was increasingly turning to alternative broadcasting options to RTÉ.

In the 1950s rural Ireland continued to empty while innovative towns and villages tried to halt their demise by establishing festivals and events to attract visitors – and to entice locals to spend money: think Puck Fair, Listowel Harvest Festival, Rose of Tralee and Galway's Oyster Festival. Many of the festivals used simple megaphones along the streets to facilitate short-term localised broadcasting for the duration of these events (and because they weren't formally broadcasting, they weren't breaking any law).

The RTÉ mobile studio at the Spring Show in Dublin's RDS, in May 1971.

These megaphone broadcasts added to the atmosphere while drumming up business through advertisements. Local competition between pubs and other businesses was an incentive to advertise. The stations took requests, even though it was understood that no one would hear them except for those out on the street or in a field (with another megaphone hoisted up on a strong post or pole like you might still see today at a sports day) where the event was being held. Requests were even played for individuals and families long emigrated to England or America. Letters from home to those in New York or Boston would recount how their name had been broadcast on Radio Listowel during the Harvest Festival or on Radio Tralee during the Rose of Tralee Festival.

But by the 1970s, as Irish-speakers expressed their right to a station in Irish (which prompted the hurried establishment of RnaG in 1972), others simply wanted to hear music, accents, news and stories that were closer to home. Eventually – harnessing its experience of mobile recording and outside broadcasting – RTÉ set up its first Community Radio Mobile Service in 1975. It spent fortnights in several towns across the country in the coming years, even piloting a countywide version in Mayo for a month

Liam Nolan, presenter of *Here and Now*, a weekday show featuring current affairs, people and the arts, with Caroline Boland, who had come first in recitation at that year's Feis Maitiú, in the RTÉ Radio temporary studio at the RDS Spring Show in May 1971.

in 1985. The service provided communities with equipment and a temporary studio, while the content and presenters were led by locals. It was the broadcaster's stated ambition to service a network of local radio. But the experience just gave people a taste of what they were missing out on and many communities set up their own pirate stations when the 'official' temporary station had left. For example, mere months after RTÉ's Community Radio Mobile Service in Mullingar in 1978, pirate station Big M went on air. And so the pirates spread across the country, giving a voice and choice to listeners previously limited to the usual RTÉ offering.

It was not technically complicated to broadcast a pirate station, and punishment, if caught in the act, was not harsh. While the number of pirate stations proliferated, the influence of some, such as South Coast Radio and Radio ERI in Cork, became significant. They captured a 'youth' audience drawn to their popular music offering, supported by local advertising revenue. And so, in 1979 – just as it had done in 1972 with Raidió na Gaeltachta – RTÉ decided that if you can't beat them, join them, launching Radio

RTÉ broadcast crew in the recently acquired mobile studio, in June 1974. From left to right: the engineer in charge of radio outside broadcasts Austin Bergin, producer Billy Wall, sound engineer Jimmy Keenan and broadcast assistant Niamh Kenny.

1970s RTÉ community radio staff with the outside broadcast van used to pilot local radio in several towns across the country when the only alternative to RTÉ Radio in the country were the pirate stations.

Local RTÉ Community Radio West project participants in Castlebar, Co. Mayo, in September 1981. Thirty-five transmissions, each of approximately three hours duration, were made.

(Opposite page) Outside broadcast van for coverage of Radio 2 Dublin City Marathon 1981.

2 on 31 May to provide a national alternative. Not long after the new station was out and about around the country with the Radio 2 Roadshow, introducing its new personalities (many who came from the pirates, including Gerry Ryan, Marty Whelan, Vincent Hanley and Dave Fanning), making several outside broadcasts and attending most major music festivals.

Through the 1980s, much larger pirate stations such as Radio Nova and Sunshine joined in the market for listeners, especially in Dublin. It would be December 1988 before all illegally run stations were taken off air and the legalised radio landscape that followed of local, regional and national community and commercial radio began to broadcast alongside RTÉ.

By the 1990s RTÉ Radio had extended its countrywide visibility with radio cars, mobile studios and roadcasters with their own satellite transmitter systems. The daily programmes took to the road, mixing a gathering of locally recorded material with live broadcasts from towns and villages. A national station, there on the ground.

RnaG reporter Seán Ó Tuairisg interviews students from Scoil Chuimsitheach Chiaráin in An Cheathrú Rua, Co. Galway, in 1972. Left to right: Bríd Ní hOichín, Treasa Ní Dhonncha, Bríd Ní Dhonncha, Caitlín Ní Gríofa, Bríd Ní Loideáin and Máire Ní Dhonncha.

RANGE ROVER
895 TZI

Broadcasters, guests and production crew during the RTÉ Community Radio West project in Ballina's Moy Hotel, Co. Mayo, in September 1981. Technical operator Benjamin Walkin is seated at a control desk.

The 2FM Roadcaster.

▼
RnaG Roadcaster, 'An Ródaí', in Clonbur, Co. Galway, in June 1995.

Everything changed in March 2020 – but RTÉ Radio, like other broadcasters, carried on. Enabling presenters to broadcast from home was one way of reducing health risks while keeping services running, so RTÉ supplied presenters with remote broadcasting kits that included microphones, headsets and, in some cases, small mixing desks.

Presenters were suddenly broadcasting live from kitchens, spare rooms and home offices, while producers and engineers supported them remotely, but it was difficult to replicate the chemistry of a live studio. Interviews and discussions lost some of the spontaneity and connection that comes from having presenters and guests in the same place.

John Bates was among those who managed these new working conditions. He remembers driving into RTÉ during the first lockdown: the roads completely empty from Phibsborough to Donnybrook, not a single car in sight. The silence was deafening, broken only by birdsong – like something out of a disaster movie. The Radio Centre was on alert in case an outright spread of Covid meant that it could not be used, so RTÉ prepared a contingency studio in Croke Park, thanks to its strong network links back to RTÉ. The GAA provided a studio space, complete with a control room and production area. The goal was to build an enhanced outside broadcast set-up that fully recreated an RTÉ radio studio. The team installed a network switch, phone

Bryan Dobson and Claire Byrne presenting on Radio 1 from home, 2020.

▲
The RTÉ Concert Orchestra recording remotely from one another for broadcast in 2020.

►
Producer Alistair McConnell in temporary Covid studio in Croke Park during *Today with Sean O'Rourke*.

and production tools, creating a like-for-like environment. *Today with Sean O'Rourke* was the first programme to broadcast from the temporary facility. The programme team, guests and staff travelled to Croke Park, with social distancing measures in place.

The Opening Night Gala concert of Lyric FM on 1 May 1999 with broadcast live from the University Concert Hall, Limerick was hosted by Mary Kennedy. It featured the RTÉ Symphony Orchestra with Alexander Annissimov (conductor) and soloists Barry Douglas (piano) and Suzanne Murphy (vocals) and opened with Aaron Copeland's *Fanfare for the Common Man*.

Being there – beyond the confines of the studio – was part of how each new station announced its arrival: in 1926, 2RN crossed Dublin to broadcast the Army Band from Beggars Bush Barracks; in 1972 RnaG carried Aifreann Uí Riada's from Carraroe church to listeners; and in 1999, lyric fm launched with a live performance of Handel's *Messiah* from the University of Limerick. Radio 2 quickly took its new presenters on the road to meet listeners where they lived and gathered. Across a century of Irish broadcasting, outside broadcasts were far more than technical exercises. They have brought radio into the heart of communities, reinforcing one of the medium's enduring strengths: the ability to make listeners be part of events as they happen.

RTÉ lyric fm Quartet performing at Bloom 2019, lyric's twentieth birthday year. The quartet of Evelyn Grant, Mary McCague, Jean Kelly and Gerry Kelly, was part of RTÉ lyric fm Outreach. They travelled around Ireland performing in schools and at local events.

APPEALING MUSIC

SPOTLIGHT

In 1984 Larry and Margaret McStay came onto Gay Byrne's morning radio show. Their baby Colin required a liver transplant. A public appeal on the programme enabled his transplant go ahead. A Gay Byrne fund went on to support many other good causes until the show ended in 1998. Brief appeals were part of Sunday programming from the 1920s until the 80s and newspaper notices and reports about them offered sometimes stark glimpses of the country.

May Fogarty was manager of the Empire Theatre in Galway when she organised the city's first live radio concert in 1927 to appeal for support for the recently bereaved families of forty-five drowned fishermen in the Cleggan disaster. Newspaper reports illustrated excitement at the new medium, as well as an understanding of how the broadcast could enhance the appeal. 'The Vast Unseen Audience' of *The Connaught Tribune*'s report would have heard Mr T. J. W. Kenny of the Galway Chamber of Commerce appeal 'to a wider audience not only in Ireland but to Irishmen beyond the seas ... how no fewer than 154 people had been left without breadwinners – some orphaned, and all left penniless, to the charity of their neighbours and the charity of the people of Ireland'. Mr Kenny told those listening at home and present in the theatre that people in the Four Provinces of Ireland Club in London were listening to the broadcast and that his son had telephoned him to say that he was hearing the concert loud and clear on their wireless at home. The other speaker was Professor Liam Ó Briain of University College Galway who spoke in Irish, acknowledging the bereaved families in their own language.

The Connacht Sentinel of 22 November noted how the volume of clapping was reflective of the large audience in the theatre. *The Connacht Tribune* of 3 December reported that every item was heard clearly in London, Cambridge, Middlesborough and 'in the six counties of Northern Ireland'. In that same issue it published a letter from 2RN station director Séamus Clandillon to May Fogarty:

> Dear Miss Fogarty, allow me to express my thorough appreciation of the beautiful programme which you were able to give us from Galway. I was particularly struck, of course, by your own magnificent artistry and technique, and I must say that the background supplied by your orchestra and the wind players from the St. Patrick's Band was excellent. The opening item by the band was very fine, and Mrs Joyce's singing was very well received indeed. The whole programme, in fact, revealed a very high standard. I was particularly glad of this because I am from Co. Galway myself.

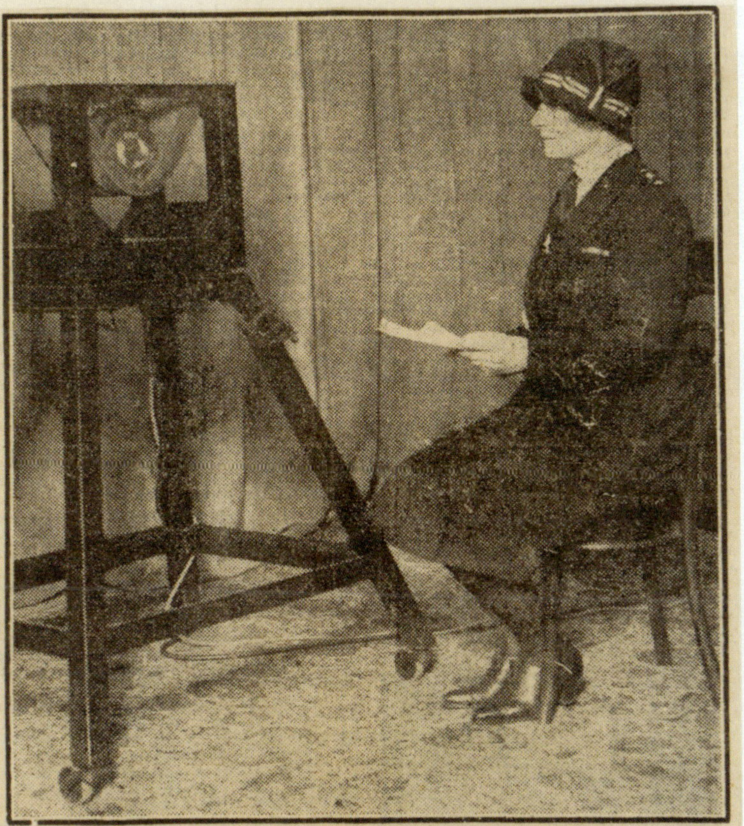

St. John Ambulance Brigade Appeal.—Miss Blandford, Divisional Lady Superintendent, broadcasting an appeal on behalf of the Brigade from 2RN. —Photo, Kelly.

The *Irish Independent* in May 1929 carried this photo of a Miss Blandford in the 2RN studio appealing on behalf of St John's Ambulance.

APPEAL IN AID OF WOMEN'S SHELTER

The annual radio appeal in aid of St. Joseph's Free Night Shelter for Women, Brickfield lane, Cork street, Dublin, will be given to-morrow night at 10.15 o'c., by the Rev. W. J. O'Connor, Oblate Fathers, Inchicore.

There were 14,071 admissions to this very charitable institution during the past year, and 28,142 free meals served. The charitable public easily can estimate the cost of such an institution and the great need for help to carry on the good work.

RADIO APPEAL

An appeal on behalf of St. Mary Magdalen's Home, High Park, Drumcondra, Dublin, will be made from Radio Eireann at 10.10 p.m. to-morrow by Very Rev. R. J. Tyndall, S.J. Superior, St. Francis Xavier's, Gardiner St., Dublin. The home is maintained by voluntary subscriptions and is in charge of the Sisters of Our Lady of Charity of Refuge.

MAGDALEN ASYLUM APPEAL

In a radio appeal for the Magdalen Asylum, Lower Gloucester Street, Dublin, Rev. Fergal McGrath, S.J., said it was under the care of the Sisters of Our Lady of Charity of Refuge and provided shelter for 150 women and girls. The Sisters received no grants from public funds, and managed to keep going on the proceeds of laundry work and voluntary subscriptions, for which they confidently asked.

LISTEN

ON TO-MORROW (SUN.) at 10.10 p.m.
To the APPEAL by
VERY REV. ROBERT TYNDALL, S.J.
On behalf of
St. Mary's Asylum, High Park
Drumcondra, Dublin
220 Magdalens rely on your generosity

APPEAL FOR ORPHANAGE

A coming-of-age appeal was made from Radio Eireann last night on behalf of St. Teresa's Home, Temple Hill, Blackrock, Co. Dublin, by District Justice McCarthy, of the Metropolitan Children's Court.

Founded in 1925, on the Feast of its Patroness, the Little Flower, the home provides for 150 orphaned motherless boys between the ages of 3 and 10, and it is altogether dependent on voluntary subscrip tions.

District Justice McCarthy said that the home carried an overdraft of £3,200, and its expenditure for the past year reached the formidable figure of £5,000.

Reports on various radio appeals in 1946, many of which were on behalf of Magdalene laundries.

Clockwise from bottom:
The *Irish Press*, 9 October 1946.
The *Irish Independent*, 7 September 1946.
The Irish Times, 31 August 1946.
The *Irish Press*, 7 September 1946.
The Irish Times, 23 September 1946.

Self-Aid: Make it Work was another benefit concert, staged in the RDS in Dublin and broadcast simultaneously on RTÉ television and Radio 2 in 1986. The cover of the *RTÉ Guide* of 16 May featured a collage of those who would perform including Elvis Costello, Christy Moore, Mary Black, Chris Rea and In Tua Nua. Inside were listed the twenty-seven acts taking part, which included U2, Van Morrison, The Boomtown Rats, Clannad, Scullion, The Pogues and Rory Gallagher. 'We are trying to get the whole country thinking about unemployment, thinking about ways of solving it that haven't been thought of before, both on an individual level and a community level,' said RTÉ's Niall Mathews, who, along with Tony Boland, was behind the event.

Like the concert in Galway in 1927, focus was given to the radio listeners (and now the television audience). It was to be a double stereophonic experience: watching on television and tuning in to FM sound on radio would be the best way to experience the event.

Self-Aid was broadcast on 17 May 1986 for 14 hours until 2 a.m. the following morning. It was inspired by Live Aid, the mammoth 1985 concert spearheaded by Bob Geldof that had raised millions for famine-stricken Ethiopia. Self-Aid had somewhat more contentious, if well intentioned, purposes. It wanted to raise consciousness of unemployment in Ireland as something everyone could contribute to fixing, raise funds to create work and to attract pledges for actual jobs. That year the country's unemployment figures reached over 17 per cent and thousands emigrated. Critics picketed RTÉ in protest that Self-Aid took the onus off government to solve unemployment. The government offered to contribute to the fund but it was rejected, exemplifying the broken state of the country and a reminder that broadcasting's attempt to mirror a sense of the times can have mixed results.

Admission ticket to Self-Aid, 1986.

Clockwise from top left: the *RTÉ Guide* of 16 May 1986, Shane McGowan (The Pogues), crowds enjoying the concert, Bob Geldof (The Boomtown Rats) and Leslie Dowdell (In Tua Nua).

THE

AGUS ANOIS

CHAPTER 7

My earliest regular contributions to radio were in Irish, on RnaG in the 1990s. I was running a theatre company, commissioning plays in Irish and the programme Ardán na nEalaíon, presented by Seán P. Ó hEalaí, had me on to talk about them. Before long I was also reviewing other plays and covering the visual arts. I came to experience radio *as Gaeilge* as integral to what RTÉ broadcasts daily, apart from more sporadic single programmes, series and Nuacht on Radio 1. *Agus tugam liom í – mar a dhéanann roinnt de mo chomhghlacaithe – isteach sna cláracha raidió a dhéanaim agus a chraoltar amach sa saol* / I bring Irish with me – as some of my colleagues do – into the programmes I make, which are broadcast out into the world.

The use of the Irish language and the variety of what was on offer in Irish to 2RN listeners from 1926 through to present-day RTÉ Radio could reasonably be said to have ebbed and flowed across its first century. As Irish radio was taking the first tentative steps in its self-invention, so too was the new state; how to forge an uncontentious, unifying identity and purpose was very much in the air. There was a great deal of talk and correspondence about what was wanted and what should be heard on the new station. There was debate about entertainment versus educational and informative content. Provision of programmes in Irish, as well as the use of it in general, tended to fall into the latter. Radio, as the State saw it, was the perfect medium for its Irish-revival ambitions. Here, after all, was a medium with voice and sound at its core. What better tool with which to promote the language, firmly placed as it was at the heart of people's homes, in their kitchens and sitting rooms?

Graph illustrating the numbers of license holders from 1926 to 1948 and published in the 1948 Radio Éireann Yearbook.

STATISTICS

LICENCES.

Licence Fee : The annual fee for a wireless receiving licence is 12/6.

Licences : The number of licences increased from 186,727 on 31st December, 1947, to 261,321 on the 31st December, 1948, an increase of 74,594.

The percentage increase was 39.9 as against an increase of 3.7 in 1947 over the 1946 figure.

The figures for the number of licences at the end of each calendar year since the inception of broadcasting in Ireland are as follows :—

TABLE I.

Year	Licences	Increase or decrease on previous year
1926	9,867	
1927	26,935	+ 17,068
1928	26,406	— 529
1929	25,733	— 673
1930	26,015	+ 282
1931	28,046	+ 2,031
1932	31,440	+ 3,394
1933	45,008	+ 13,568
1934	60,192	+ 15,184
1935	78,627	+ 18,435
1936	98,949	+ 20,322
1937	112,192	+ 13,243
1938	148,811	+ 36,619
1939	166,275	+ 17,464
1940	179,563	+ 13,288
1941	183,303	+ 3,740
1942	169,933*	— 13,370
1943	169,372	— 561
1944	171,105	+ 1,733
1945	172,705	+ 1,600
1946	180,005	+ 7,300
1947	186,727	+ 6,722
1948	261,321	+ 74,594

* *Drop caused by shortage of batteries during war years.*

TABLE II.

The following table of Ratios shows the distribution of Licences for the year 1948 in relation to the population.

Per 1,000 of Population.
(Based on 1946 Census of Population.)

LEINSTER	.. ..	119 or 1 in 9.
MUNSTER	.. ..	79 or 1 in 12.
ULSTER (3 Counties)	..	46 or 1 in 21.
CONNACHT	.. ..	51 or 1 in 19.
GENERAL DISTRIBUTION	..	88 or 1 in 12.

38

However, in the early days wireless owners – not to mention licence holders – were scarce, so the station was unlikely to contribute significantly to the language revival project, at least not in the short term. In 1925 the title and status of An Ghaeltacht had been granted to areas in which Irish was deemed to be the dominant daily spoken language. Then on 18 April 1926, only months after 2RN went on air, Ireland held the first census as an independent country. Results confirmed that only some 18 per cent of the population at the time of 2,971,992 declared themselves to be Irish speakers.

Irish was the first language spoken on air on the opening evening of 2RN. Séamus Hughes, referred to as Séamus Mac Aoidh in official correspondences relating to his role as the station's continuity announcer, spoke those words. He began: 'Sé seo Raidió Bhaile Átha Cliath ag glaodhach; This is 2RN, the Dublin broadcasting station, calling ...' Although the evening's invited guest speaker, Dr Douglas Hyde, began in English, within two sentences he continued with, 'I will now continue in our own language ...' and off he went in Irish for the remainder of the broadcast. Hyde was a Gaelic scholar, playwright and founder of Conradh na Gaeilge (the Gaelic League), the grass-roots organisation founded to re-establish the everyday use of Irish. In 1938 he would become the first President of Ireland. During that opening night of Irish radio, he spoke about the importance of cultivating the language because, he said, 'a nation is made from inside itself. It is made first of all by its language.'

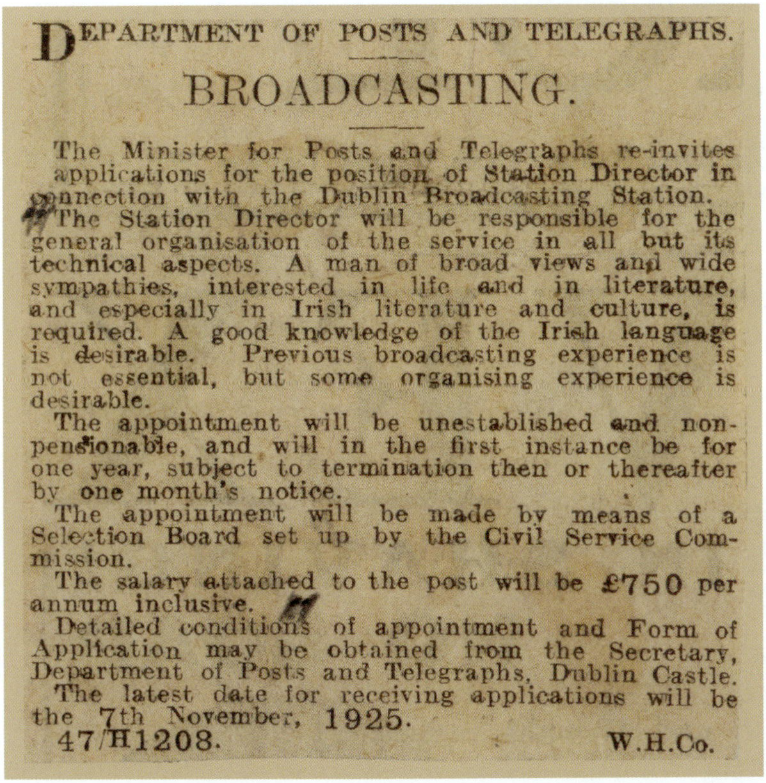

DEPARTMENT OF POSTS AND TELEGRAPHS.

BROADCASTING.

The Minister for Posts and Telegraphs re-invites applications for the position of Station Director in connection with the Dublin Broadcasting Station.

The Station Director will be responsible for the general organisation of the service in all but its technical aspects. A man of broad views and wide sympathies, interested in life and in literature, and especially in Irish literature and culture, is required. A good knowledge of the Irish language is desirable. Previous broadcasting experience is not essential, but some organising experience is desirable.

The appointment will be unestablished and non-pensionable, and will in the first instance be for one year, subject to termination then or thereafter by one month's notice.

The appointment will be made by means of a Selection Board set up by the Civil Service Commission.

The salary attached to the post will be £750 per annum inclusive.

Detailed conditions of appointment and Form of Application may be obtained from the Secretary, Department of Posts and Telegraphs, Dublin Castle.

The latest date for receiving applications will be the 7th November, 1925.

47/H1208. W.H.Co.

Advertisement for the position of the first director of 2RN.

Translation attached

SAORSTÁT ÉIREANN

Department of Posts and Telegraphs.

Application for position as Stiúrthóir.................

in connection with Broadcasting Service.

(Latest date for receiving applications - 12th November 1925)

---

1. Name (in full) ...Séamus de Clanndiolúin.. Seamus Clandillon.
2. Postal address .Tig an Custuim, Luimneach 7 17 bóthar Droma Cuilinn, Droimconnrach.
3. Date of birth ......6. 6. 1878...........
4. Place of birth of (a) applicant .Loch Cútra, Gort inse Guaire, Co na Gaillimhe
   (b) applicant's father Áth Í, Condae Cill Dara.
5. State academic qualifications .B.A. le honóir i N.U.I. (BA. Hons NUI)
6. State other professional or technical qualifications, or any special qualifications for post now sought ....
....Amhránaidhe Gaedhealach, ar feadh na bliadhanta, ag Cuirmibh Mór Ceoil i...
....nÉirinn, i nAlbain 7 i Sasana... Moltóir ceoil ag an Oireachtas, ag......
....Feiseannaibh ar fud na hÉireann 7 ag Feis Tailteann. Do chuireas ar bun..
.....Feiseanna, Cuirm Ceoil 7 Taisbeántais [illegible] ar fud na hÉireann...7c....
7. Knowledge of Irish ...Tá an Ghaedhilg ar fheabhas agam.
8. Previous experience (name and address of employer with dates and nature of appointment to be given).
....Scoil an Ceard-oideachais, Cluain Meala (dá bhliadhain) 1903-1905.........
.....Scoil na Teicniúdeachta, Clanna Caoilte..1905-1912.......................
9. State particulars of service in any Government Department, and whether you have a pension in respect of such service .Féach tuas [illegible]
..[illegible] ó 1912 im Cigire (7 Cigire Reanna) fé Coimisiún an Oideachais Náisiúnta. Tá ceart pinsin agam.
10. Service (if any) with National Army Dá bhliadhain go leith i bPortobello (Dependants' Claims Dept. Army Finance)
11. Are you in good health and free from physical defect? Táim. Moladh le Dia.
12. Are you free from pecuniary embarrassment? Táim..................
13. Salary of present or last position £.550 + bonus..................
14. Whether at present in employment ...Táim, fé Coimisinéirí an Oideachais Náisiúnta
15. References (who must be reputable persons not related to applicant. The present or last employer (if any) should be one of the references)
.......Sir Seórsa MacFloinn...Cathaoirleach, Coimisinéirí an Oideachais Náisiúnta
.......Piaras Ceannt... leas-Rúnaidhe.........................
.......C. H. Darley...Ard-Cigire..................................

Signature of Applicant Séamus de Clanndiolúin...

Date ...10. 11. 1925...

T.V. 1104

A f[ee of ...] [Civi]l Service Stamps, which can be obtained at any Head [...] s[...] accompany the application.

Stampaí anseo

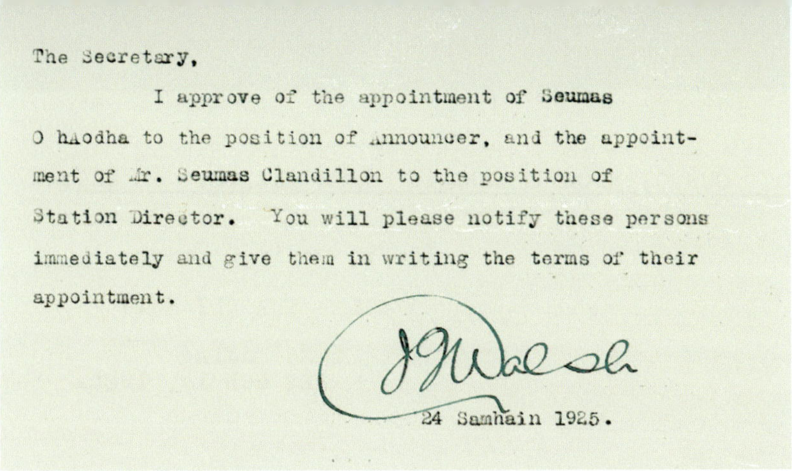

The Secretary,

I approve of the appointment of Seumas O hAodha to the position of Announcer, and the appointment of Mr. Seumas Clandillon to the position of Station Director. You will please notify these persons immediately and give them in writing the terms of their appointment.

J J Walsh

24 Samhain 1925.

The letter from J. J. Walsh, minister for posts and telegraphs, approving of the positions of Séamus Hughes as the station's announcer and of Séamus Clandillon as director of the station.

(Opposite page) The application made by Séamus Clandillon for the position of director of the new station.

The new station director was Séamus Clandillon. He was a devoted Irish-speaking Galwegian and respected traditional musician, singer and song-collector. He was also a civil servant seconded for the role from the department of health, having applied for the station director job by public competition. Like many people at the time, he used the Irish version of his name, Séamus Ó Chlanndilliún, for his signature, though people mainly referred to him as Séamus Clandillon.

Songs in Irish were included in the opening programme, with Clandillon himself singing 'An Goirtín Eornán', 'Bríd Gheal Bhán' and 'Cnocáinín Aoibhinn Chill Mhuire'. The latter he later recorded commercially – it can still be heard today. The songs immediately followed the No. 1 Army Band's 'First Irish Fantasia', directed by Colonel Fitz Brace after Hyde's opening speech. Other songs in Irish on the night were 'An Beinsín Luachra', 'Cois Abha Móire' and 'Caoine Chill Chais' sung by Clandillon's wife, the singer Máighréad Ní Annagáin.

Each day one of the announcer's duties was to provide a *Tuairisc an Bhollscaoire* or Announcer's Report of what went out on air. From 13 January 1926 the daily Announcer's Report ends with the words *Deireadh gnó* (End of business), a nod towards Irish featuring in station administration (although there is no fada in the spelling of *gno*, nor are fadas used consistently in the Announcer's Report). The first *Ceachta Gaedhilge* (Irish lessons) opened the broadcast of 25 January and was given by Seán Ó Conchubhair. Seán Ó Cuiv delivered *Scéilíní Beaga* (Little Stories), an item that ran for twelve minutes on 27 January. On 30 January M. O Lionain provided a fifteen-minute talk called *Stair Clóbhuailte* (A History of Printing).

STAISIUN CHRAOIBHSCAOILEACHAIN 'ATH CLIATH. - 2RN.

36-39, Sraid Denmarc,

BAILE ATHA CLIATH.

Mi Na Nodlag
22 adh la, 1925.

Do'n Chraoibhin Aoibhinn,
An Dochtiuir Dubhglas de h-Ide,
Colaiste na h-Oll Sgoile,
BAILE ATHA CLIATH.

Dom sheah-charaid, An Craoibhin, e seo,

Seo athchuinge ort, a Chraoibhin, o Aire Phuist agus Telegrafa, a d'iarraidh ort theacht annso go dti an Studio chun an Staisiun so d'fhosgailt duinn ar la chinn bliadhna.

'Se an rud a bheidh uainn na oraid gairid , brioghmhar, do'n saghas is col duit fhein,(oraid i nGaedhilg ta's agat),a sheasochaidh ar feadh cuig noimeatai deag, no deich noimeatai ar a laighead; a radh le muinntear an Domhain mhoir go bhfuilimid, na Gaedheala, ag seasamh ar ar mbonnaibh aris, agus chun a theasbaint go bhfuilimid i measg na Naisiun uilig.

Ceathramhadh chun a h-ocht sa trathnona caithfidh an caint a thosnughadh, agus ma bhionn tu sa Studio roimh re is amhlaidh is fearr e chun taithighe a dheanamh ar an nGleas.

'Se mian Aire agus Runaidhe Phuist agus Telegrafa, go mbeidhtha-sa againn thar aon fhear i nEirinn, agus dar ndoigh, 'se mo mhian fhein e leis.

Mise, le h-ard mheas,

S T I U R T H O I R.

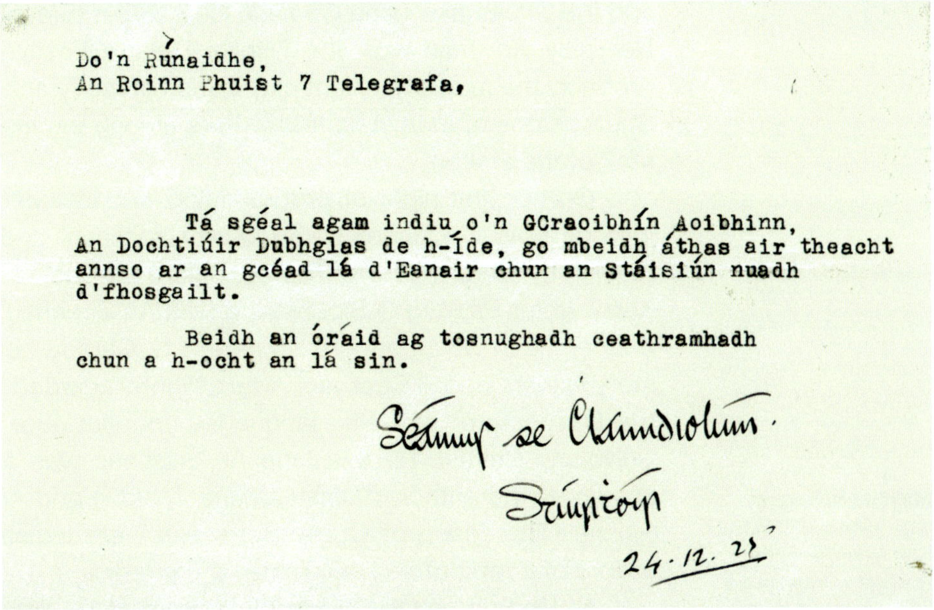

Do'n Rúnaidhe,
An Roinn Phuist 7 Telegrafa,

Tá sgéal agam indiu o'n GCraoibhín Aoibhinn, An Dochtúir Dubhglas de h-Íde, go mbeidh áthas air theacht annso ar an gcéad lá d'Eanair chun an Stáisiún nuadh d'fhosgailt.

Beidh an óráid ag tosnughadh ceathramhadh chun a h-ocht an lá sin.

Séamus de Clanndiolúin.
Stiúrthóir
24.12.25

Using *cúpla focal* – a few words in Irish – reflected the station's official language and its operation under a government department and was based on the belief that people would come to know Irish through daily exposure. Staff were employed to write material in Irish as well as English. Among them was Máiréad Ní Ghráda, who joined the station in 1927. Ní Ghráda was bilingual from childhood. Reared in Co. Clare, she won a scholarship to UCD, where she took a BA in Irish, French and English, followed by an MA on *An Aoir sa Ghaeilge* (Satire in Irish), supervised by Dr Douglas Hyde, and was publishing articles and stories in literary reviews from 1919. She taught and was secretary to Ernest Blythe, minister for finance, as well as being involved in Irish language theatre before she joined the station. She spent a couple of years looking after programmes for children and women before being appointed principal announcer. She was in fact the first woman announcer on these islands; the BBC would only follow some years later. She would also occasionally stand in for the regular announcer in giving the station's Irish lessons. Her plays were staged in the Peacock and Gate theatres; she translated several international texts into Irish as well as many of the famous Penguin Ladybird children's books, including *Lúidín Ó Laoi* (Rumpelstiltskin), *Micilín agus an Dá Lepreachán* (The Elves

(Opposite page)
Letter from Director Séamus Clandillon, inviting Dr Douglas Hyde to open the station on New Year's Day 1926. He addresses the letter to his old friend, An Chraoibhin, as Hyde was also called.

(Above)
Letter from Director Séamus Clandillon to the department of posts and telegraphs letting them know that he got word from Dr Douglas Hyde that he is delighted to come to open the station.

and the Shoemaker) and *Codladh Céad Bliain* (Sleeping Beauty). Her play *An Triail* won the Dublin Theatre Festival in 1966 – probably the most important play in Irish of the twentieth century. She was one of many talented writers among the Irish-speaking staff of the station.

On her first night of programming, she included a story in Irish by F. W. O'Connell (aka Conall Cearnach) and a song in English by Gerard Crofts. The next night she included the west Kerry writer Pádraig Ó Siochradha (aka An Seabhac), who read a story. Jean Nolan sang a song and Ní Ghráda herself told a story in Irish. Conall Cearnach was a highly regarded linguist and the first lecturer of Celtic languages and literature in Queen's University Belfast. By this time An Seabhac was already well known as the author of *Jimín Mháire Thaidhg* and *An Baile Seo 'Gainne* (This Place of Ours). Both men were indicative of the calibre of contributors keen to be on the radio.

Ní Ghráda usually divided her programme time into three sections: *Scéal i nGaeilge, amhrán nó ceol i lár baill agus scéal i mBéarla*; A story in Irish, a song or music in the middle and a story in English.

Máiréad Ní Ghráda, who joined 2RN in 1927. She would become the first woman radio announcer on these islands, as well as a celebrated playwright, short story writer and translator.

Ní Ghráda recalled the station's daily opening announcement was 'The Dublin station calling Dublin 2RN', and that it was Séamus Clandillon who thought of using it. She didn't think many people understood what the name was saying, and that this was probably because it was not pronounced as 'to Erin'. When the station acquired its new name of Athlone or Raidió Áth Luain in 1932 (in recognition of the new radio transmitter there), Ní Ghráda remembered using the Irish name on air. When the station's name changed again in 1937 the announcers adapted to calling it Radio Éireann, using the Irish pronunciation 'Rah-dio' Éireann. Ní Ghráda praised Clandillon for insisting on the voice of the station being fully Irish on St Patrick's Day while he was director.

The Gorey (Wexford) Gaelic League Players who broadcast the Irish play, "An Feall," from 2RN last night. Giolla Criost O Broin, the author, is seated in centre of front row.
[I.P. Photo.

Many groups performed live readings and performances in Irish in the early years of 2RN. A photograph from *The Irish Press* of 25 September 1935 shows the Gorey, Co. Wexford Gaelic League Players with the writer Giolla Criost Ó Broin, whose play *An Feall* they performed on the station.

A headline in *The Irish Press* of 17 April 1935 illustrates the daily presence of the Irish language on the radio: '*Gael-Nuacht O Radio Ath-Luain*' (Gaelic News from Athlone Radio) about the station's programme, called *Scáthán na nGaedhal* (Mirror of the Gael). It was a feature news programme that ran four nights a week. The newspaper reported that the previous evening the programme had given a countrywide round-up of activities of Gaelic League societies in Donegal, Limerick, Galway, Mayo, Cork and Wicklow.

A photograph of Pádraig Ó Móráin in *The Irish Press* of 1 November 1935 in front of the microphone in the GPO when broadcasting live for *Scáthán na nGaedhal*, the Irish language news programme.

The Limerick society was in the throes of rehearsing the plays *An Dochtúir Bréige* (The Fake Doctor) by Fionán Ó Loinsigh and *Cluiche Cártaí* (Card Game) by Piaras Béaslaí; Fáinnes (gold, ring-shaped brooches) for speaking Irish were presented to three members of the Béal an Áth Gaelic League society, while Gardaí, newly stationed in Killorglin, were said to be taking great interest in the revival of Irish and helping the League. Looking forward to that evening's broadcast the article went on to say how, 'In the Athlone programme tonight Cú Uladh, President of the Gaelic League, will speak in Irish about the history of the Irish tongue. Later in the evening the Cómhar Drámaíochta will produce a Gaelic play entitled *Ruaidhrí Réamh-Ráite* [Rory aforementioned].'

The near-daily coverage in the newspapers of what was broadcast illustrated the public interest in radio, as well as the interest of the print media. Radio Éireann did not have its own radio journal until the *RTV Guide* began publishing weekly at the end of 1961, a month before RTÉ Television began.

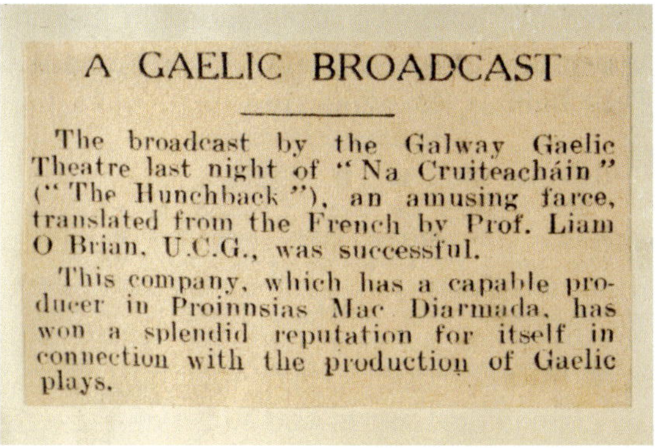

A GAELIC BROADCAST

The broadcast by the Galway Gaelic Theatre last night of "Na Cruiteacháin" ("The Hunchback"), an amusing farce, translated from the French by Prof. Liam O Brian, U.C.G., was successful.

This company, which has a capable producer in Proinnsias Mac Diarmada, has won a splendid reputation for itself in connection with the production of Gaelic plays.

Mr. and Mrs. Seumas Clandillon.

From the 1930s *Scáthán na nGaedhal* reported recent events and was broadcast four nights a week. The programme was often referred to in Dáil debates. (Much attention was also focused on pronunciations and accents – *Scáthán na nGaedhal* was considered a tool for learning Irish rather than a dedicated news and information provision for those whose first language was Irish). By 1938 *Nuacht an Lae* (News of the Day) was broadcast nightly, while news in English was broadcast twice a day. For a period in the 1940s each newsreader with a different Irish dialect would be rostered for a week in rotation to include as many Irish dialects as possible. But this did not stop suggestions coming in from the Dáil that there was no better place to learn the language than from *An Nuacht*. It was even suggested that difficult words could be explained to listeners before the broadcast of *An Nuacht*.

In 1944 Radio Éireann introduced a weekly series called *Is Your Irish Rusty?* Sorcha Ní Ghuairim, a native speaker from south-west Connemara, introduced a refresher course for people who had once learned Irish but then forgotten most of it. The very successful *Listen and Learn* series started in 1945, with listenership across Ireland and beyond to Britain and France. It even competed with Sunday evening's *Question Time* as the station's most popular programme. *Listen and Learn* books sold extremely well: nearly 30,000 for the first series and 11,000 more for the second. The programme was so popular that HMV

(Left)
A report of a broadcast of the Irish-language play *Na Cruiteacháin*, relayed from Galway.

(Right)
A photograph of Séamus Clandillon and his wife, singer Máiréad Ní Annagáin, that accompanied a feature on them in the September 1930 edition of *Model Housekeeping*.

released an LP record of the texts in Irish included in the programmes. In 1947 a new series, *Conas Adéarfa?* (How Would You Say?) began broadcasting, aimed at helping more accomplished speakers to improve their fluency.

The Radio Éireann Yearbook was published for the first time in 1946, documenting what was broadcast during the year, including sports, talks and religious services, as well as Irish-language programmes. These featured excerpts from *An Béal Bocht* (The Poor Mouth) by Myles na gCopaleen, as well as the short stories 'An Beo' (The Living) by Llam Ó Flaithearta and 'An Bhean Lán de Stuaim' (The Wily Woman) by Seosamh Mac Grianna.

The bestselling *Listen and Learn* books that accompanied the very popular Irish-language learning programmes broadcast on Radio Éireann in the mid-1940s.

The introduction to the 1946 Radió Éireann Annual Report by the station's director, Séamus Ó Braonáin, is only in Irish. The statement being in Irish and the use by the director of his name in Irish reflects the common use of Irish on official documents, sometimes to the exclusion of English. It is not to be taken as an accurate reflection of the actual use of Irish by the individual. Director Séamus Ó Braonáin was in fact known to his colleagues and friends as Séamus Brennan. An earlier station director, Dr T. J. Kiernan, did once say 'you could get away with murder in Irish' on the air. Kiernan was referring to the fact that listeners tended to comment on the use of the language itself rather than what the station was presenting in Irish to listeners.

The series and programmes listed indicate that the types of programmes being offered in Irish tended towards those offered in English. A series called *Ceist agam Ort* (Answer me this) was similar to the station's popular English *Question Time*. There was a debate on *An Leigfear an Ghaeilge chun Báis?* (Will Irish be let die?) between Earnán de Blaghd, then managing director of the Abbey Theatre, poet and fiction writer Séamus Ó Néill and playwright Annraoi Ó Liatháin. Books were reviewed in the form of scripted talks and chaired by Peadar Ó Dubhda (who also presented the station's children's programme in Irish) along with poet and diplomat Valentin Iremonger, Dáithí Ó hUaithne, professor of Irish at Trinity College, and writer and journalist Ciarán Ó Nualláin.

Recording by Radio Éireann programme makers in Irish (as well as in English) in various dialects and accents from around the country from the late 1940s extended the types of

The Irish Press of 21 September 1946 reports that, in his second broadcast appeal for assistance with saving the wheat harvest in 1946, Taoiseach Éamon de Valera turned to Irish to say that it was still possible together to save it so that bread rationing would not be necessary.

Agus Anois, a Ghaeilgeoiri !

AG críochnú, dubhairt an Taoiseach: Agus anois, a Ghaeilgeoirí. Ba thógáil chroí dhúinn go léir an comhar agus an cur-le-chéile a chonnaiceamar ar fud na tíre ó labhair mé libh cheana.

Bhí sean-chomharsanúlacht na nGael agus aontacht íontach na ndaoine le feiceál arís i ngoirt arúir na hEireann.

Bímís buíoch do Dhia as an deis a thug sé dhúinn leis an gCríostaíocht — grá dár gcomharsain—a chleachtadh sa gcaoi sin.

Mar adeir an sean-fhocal, is maith le Dia féin cabhair a fháil. Beidh Sé buíoch ach leanacht den tréan-iarracht atá ar siúl again'n le cupla seachtain anuas. Níl ach seachtain nó dhó eile fagtha.

Déanaimís, gach duine againn, gach a bhféadfaimid sna leatheanta atá romhain'n. Gach duine ata i ndon lámh chúnta a thabhairt leis an arúr, tugadh sé é go fonnmhar agus tugadh sé é ga'n mhoill.

Tá breith fós ar an bhfómhar a shábháil má cuir tear an neart san iarracht. An cur-le-chéile, agus rath Dé ina theanta, Go raibh againn. Go gcumhdaí Dia sibh.

Gael-Linn's sponsored music programme on Radio Éireann/Radio 1 from 1953 until 1980. From left, Roibeárd Mac Góráin, Dónal Ó Móráin, Pádraic Ó Raghallaigh, presenter and Clár Ní Dhuigneáin.

programmes being made and broadcast (a development covered in the chapter relating to outside broadcasting). When the Radio Éireann Players repertory company (RÉP) was established at the same time, including fluent Irish speakers, it assisted greatly in the provision of voices being available not just for plays but for documentary programmes and readings in the same way as the other RÉP actors were available to do in English.

Gael-Linn, the organisation established in 1953 with a mandate to establish Irish as a living, thriving language, sponsored the radio programme of the Irish-language organisation, An Comhchaidreamh. It was called *Clár an Chomhchaidrimh* (The Association's Programme). Gael-Linn also ran a hugely popular football pools called *Na Línte* (The Lines). The pools were sold weekly door to door around Dublin and other urban centres around the country. *Clár an Chomchaidrimh* also promoted the key Irish language literary magazine *Comhar*. It is still thriving and has published most of the significant writers in Irish, under an equally impressive series of editors since it was established in 1942.

Clár an Chomhchaidrimh's first presenter was Pádraic Ó Raghallaigh; listeners were drawn to the natural warmth and assuredness of his voice and the fluency and clarity of his Irish. The programme was broadcast from 8.30 to 8.45 a.m. every Monday and again at 2.30 p.m., and later brought Irish music and singing into people's homes. In 1964 Donncha Ó Míodhcháin took over as presenter and Breandán Ó Dúill, himself a popular singer, was the regular presenter until RTÉ phased out sponsored radio programmes.

SOUND

Thursday

JULY 5

Athlone, 530 m. (566 k/cs.); Dublin and Cork: 240 m. (1,250 k/cs.).

Radio Éireann

8. 0 **WEATHER and NEWS**

8.10 **WEATHER**

8.15 **BOLANDS**
Movie Music from current films. Niall Boden.

8.30 **MUSGRAVE BROTHERS**
Music and Song, with Jimmy Shand and his band and the George Mitchell Minstrels.

8.45 **PHILIPS**
A light-hearted comedy programme with the accent on satire.

9. 0 **NEWS**
Weather
followed by

MORNING MELODY
Musical Miscellany compered by Ronnie Walsh and Larry Grogan.

10. 0 **NEWS**
Signpost
Weather
Closedown

12.55 **NEWS HEADLINES**
Weather

1. 0 **FRY CADBURY**
presents another episode of The Kennedys of Castleross.

1.15 **PRESCOTTS**
Music and stories from the ballet.
Compere: Ronnie Walsh.

1.30 **NEWS**
Topical Talk

1.45 **BIRDS**
Disc programme featuring Opera Corner.

2. 0 **BRADMOLA**
Two O'Clock Opera
Stories from the Operas told by Ronnie Walsh.

2.15 **LUCOZADE**
Harry Thuillier conducts interviews in hospitals.

2.30 **KATE'S COTTAGE JAM**
A new music and interview programme.
Compere: Noel Andrews

2.45 **STAGECOACH, STIRRUP AND STEER**
Ballads of the old West presented by Proinsias O Conluain.

3. 0 **NEWS HEADLINES**
Closedown

5. 0 **NEWS HEADLINES**

5. 1 **COGAR, A LEANAI**
le Peadar O Dubhda.

5.15 **BANNA CAIRDIN SCOIL UI CONAILL**
á stiúrdah ag an mBráthair O Murchú.

5.30 **GREAT EXPECTATIONS**
Charles Dickens's story of Pip's adventures.

JOHN SKEHAN
presents *Laughing For Living* tonight at 6.45 R.E.

produced by P. J. O'Connor.
Episode 2—A Guardian for Mr. Pip.
(RE recording).

5.55 **WEATHER**

6. 0 **THE ANGELUS**

6. 1 **NUACHT**
Cáint na Deardaoine

6.20 Racing Results
Announcements
Stock Exchange Prices
Signpost

6.25 **HERE'S THE DOCTOR**
Weekly Health Talk.

6.30 **NEWS**

6.45 **LAUGHING FOR A LIVING**
John Skehan presents comedy on records.

7. 0 **THE DUBLIN HARPSICHORD ENSEMBLE**
with Ruth Maher, contralto.
PURCELL: *The Golden Sonata*
PURCELL: *5 songs from the Fairy Queen.*
ARNE: Trio Sonata.

7.30 **BETWEEN OURSELVES**
A magazine for women, presented by Petronella O'Flanagan.
This week's items:
Nancy Riordan talks about rambling for profit, and in particular free foods to be gathered by the wayside.
Voters: Judith Rosenfield, a Liberal Party candidate in the recent Six Countes election, recounts her canvassing adventures.
D. M. Webster gives her her impressions of America as seen through the eyes of an Irish visitor.
Songs My Mother Taught Me: Sheila Byrne describes the repertoire of racing rollicking ballads acquired at

8. 0 **NEWS HEADLINES**

8. 1 **SALLYPARK**
A novel by Margaret Hassett dramatised for radio in six episodes by Frances McHugh. Producer: Michael Garvey.
Episode 2—Mrs. Warmbath at Sallypark.

8.30 **RANCH HOUSE REVELS**
Hilly-Billy variety with Hal Roach, Rose Tynan, Florence Lynch, Derry Power, Peter Dix, Willie Brady, Norman Metcalfe and the Mountaineers.
Written by Dermot Doolan
Producer: Joan D'Alton.

9. 0 **NEWS HEADLINES**

9. 1 **OBAIR IDIR LAMHA**
Udair aitheantúla ag caint ar an saothar atá a scríobh acu.
Dónal Mac Amhlaigh.

9.15 **MELODY RENDEZVOUS**
Presented by The Radio Eireann Singers
Conductor: Hans Waldemar Rosenor, with Tommy Dando at the organ.
Soloists: Bernadette Greavy (contralto) and Jimmy Nolan (accordeon).

9.45 **GLOR NA MUMHAN**
Caint agus ceol ó Phádraig Tyers agus Seán O Sé, amhránaí.
Uimh. 2: Amhráin ó Bhaile Mhuirne.

10. 0 **NUACHT**

10.15 **NEWS**

10.30 **TODAY IN THE DAIL**

10.45 **THE S A B C LIGHT ORCHESTRA**
Conducted by Jos. Cleber
A South African Broadcasting Corporation Transcription.

11. 0 **SINGER'S CHOICE**
Angela Carroll (contralto.)
in songs by Debussy and Poulenc.

11.15 **SOUTH SEA ISLAND MAGIC**
Music in Hawaiian and Latin - American style, featuring the Sammy Mitchell Quintet and the Melotones, introduced by Jack McGeagh.

11.30 **NEWS HEADLINES**
Late Sports Results
Weather
Closedown

Author at Work

Donal Mac Amhlaigh is featured in *Obair Idir Lámha* tonight at 9.1: R.E.

Writer Donal Mac Amhlaigh, author of *Dialann Deoraí* (translated by Valentin Iremonger as *An Irish Navvy: The Diary of an Exile*), was the subject of the Radio Éireann programme *Obair idir Lámha*, which focused on the author's current writing. He became a regular contributor from London to RnaG.

Gael-Linn began recording Irish-language songs and music on acetate discs, later adopting new recording technologies as they emerged. The first release was the song 'Bean Pháidín'. Among those recorded were the Connemara singers Seán Ó Dhonnachadh and Seosamh Ó hÉanaigh. With characteristic pragmatism, Gael-Linn also accommodated broader musical tastes, producing recordings of the Radio Éireann Light Orchestra accompanying the tenor Tomás Ó Súilleabháin, with Seán Ó Riada at the piano. Ó Riada arranged the songs according to classical musical and vocal conventions.

By the late 1960s, a rite of passage had taken hold among Irish teenagers: the summer Irish college. For three weeks – sometimes longer – students were sent to the Gaeltacht to immerse themselves in the language. This experience in its modern form, built on initiatives from the Gaelic League, had been offered to adults during the early years of the revival of Irish at the turn of the twentieth century. The Irish college experience brought teenagers together from all over the country who might otherwise not have mixed with one another in daily life. Friendships were born and letters were exchanged after the teenagers returned home.

The summer radio series *Tráth Ceist Ó Choláistí Samhraidh* (Summer Colleges' Quiz) was a series of quizzes between different summer colleges around the country. Doireann Ní Bhriain was quizmaster and interviewed students in the series, which was produced by Séamus Breathnach. He explained to the *RTÉ Guide* how, 'out of each college we chose six competitors, not only for their ability to answer questions in Irish but also for their talents. Each member of the team must be able to sing a song or recite a poem in Irish. There are prizes for the first three and consolation prizes for the others.'

Irish speakers being provided with their own radio service – especially for the recently established Gaeltacht areas – was mooted as early as 1925, a year before 2RN opened, but had not advanced much. By the late 1960s, international agitation for societal change and civil rights galvanised residents of the Gaeltacht areas to demand what they had been promised decades ago and what they saw was rightfully theirs. Gluaiseacht Chearta Sibhialta na Gaeltachta (GCSG), the

Students participating in the Saturday radio quiz, Tráth Ceist Ó Choláistí Samhraidh: *this week Coláiste Thrá Bhán, Connemara is the venue.*

QUESTIONS FOR THE GAELTACHT

IN THE mid-Sixties Irish Summer colleges began to mushroom in the Gaeltachts, their establishment being closely linked with the introduction of an oral Irish examination in the Leaving Certificate. Whatever the reason for their emergence, young people from all over Ireland went on holidays for a month in the Gaeltachts, and the habit has clung so that these colleges now provide inexpensive holidays for young city people, giving them a taste of life in an Irish-speaking area and time to enjoy themselves.

A few years back, when the Minister for Education abolished "compulsory Irish" in the Leaving Certificate, there were fears that the number of children attending the Irish Summer colleges would fall off. But this has not happened, and the numbers going to the colleges have, if anything, increased. That so many parents are interested enough in Irish to send their children to the Gaeltachts is surely an encouraging sign for the future of the language.

Last Summer, presenter Doireann Ní Bhriain and producer Séamus Breathnach visited twelve of the Irish Summer colleges, recording a quiz series in each one they visited. The twelve programmes were broadcast on RTE Radio in July and August under the title *Tráth Ceist O Choláistí Samhraidh.* Exactly the same formula is being used for this year's programme. During the twelve weeks, boys and girls attending Summer colleges in Connemara, Clare, Kerry, Donegal and West Cork participate in the programmes.

Saturday's edition comes from Coláiste Trá Bhán in Connemara; two boys and four girls take part in it. They are Gearóid O Seannain from Co. Clare, Clodagh Ní Mhathúna from Dublin, Máire and Síle Ní Dhómhnaill from Westport, Bernard O Saomain from Tipperary and Róisín Ní Chonghaile from Ballina. Each of them has to answer two-mark, four-mark and six-mark questions and either sing a song or play some musical instrument.

Listening to some of the recordings brought back by Séamus Breathnach, one is impressed by the fluency of the children in Irish, their enthusiastic responses to Doireann's questions and the way they sing their songs.

movement for civil rights for the Gaeltacht, was established in March 1969. One of the movement's aims was to achieve a 'Stáisiún raidió do Ghaeilgeoirí na tíre ar fad sa Ghaeltacht' – a radio station for all the country's Irish speakers in the Gaeltacht. Two members of GCSG, Piaras Ó Gaora (who would become news editor of RnaG) and Seosamh Ó Cuaig (who would also become an RnaG broadcaster), assisted technologically by University College Galway's engineering undergraduate Micheál Ó hÉalaithe, set up the pirate station Saor Raidió Chonamara. Unable to secure a premises, they eventually set up their studio in a rented caravan in Rosmuc. They intentionally located themselves close to the cottage that Pádraig Mac Piarais had owned and in which he had completed much of his best-known writing. It went on air at 6 p.m. on Easter Saturday, 28 March 1970.

Tráth Ceist Ó Choláistí Samhraidh **was a radio series broadcast from the Gaeltacht Summer Schools around the country in 1975, hosted by Doireann Ní Bhriain.**

Radio na Gaeltachta

Ceannaire Radio na Gaeltacha, Padraic O Raghailligh.

Uachtaran na hEireann, Eamon de Valera.

Domhnach Cásca

3.00 TUSU OIFIGIUIL RADIO NA GAELTACHTA
Dileagra na h·oscailte dá chraoladh ag **ASholise, Uachtarán na hÉireann, Eamon de Valera.**

3.10 AIFREANN UI RIADA
dá chraoladh ó Theach Pobail Mhic Dara ar an gCeathrúin Rua i láthair **A Ghráss Ard-Easborg Thuama, An Dr. Ró-Oirmh. Seosamh Ó Cuinneán.**
Coimhcheiliúraithe:
An **tAth. Tomás Ó Concheanainn** (An Cheathrú Rua)
An **tAth. Máirtín Lang** (An Ceathrú Rua)
An **tAth. Pádraig Ó Héalai** (Corcha Dhuibhne)
An **tAth. Eoghan Ó Frighil** (Anagaire)
An **tAth. Tomás Ó Murchú** (Oileán Cléire)
An **tAth. Donncha Ó Conchúir** (Cúil Aodha).
An Eipistil dá léamh ag an **gCanónach Ró-Urramach Coslett Ó Cuinn**, Uachtarán an Oireachtais, 1972.
Urnaithe dá léamh ag Ceannaire is ag Clár-Reachtairi Radio na Gaeltachta.
Cóir as Conamara, Cúil Aodha is Tamhlacht dá stiúrú ag an **Ath. Mac Fhloinn** (An Spidéal), agus **Pól Ó Corbáin, Audrey Bean Ui Chorbáin** i mbun an orgáin.
An searmanas dá stiúrú ag an **Ath. Ó Fatharta** (Tír an Fhiaidh). Tráchtaireacht ón **Ath. Tomás Mac Aodha** (Gearr Bhaile).

4.00 TIOMCHUAIRT NA nGAELTACHTAI
Cuid a hAon.
Clár irise, faisnéise, ceol is reacaireachta a cuireadh i dtoil a chéile sna ceantair éagsúla Gaeltachta.
An Fhoireann a d'eagruigh é: **Micheál Ó Sé Breandán Feirtéir, Seán Ó Tuairisc, Micheál Ó Conaola, Máirtín Ó Fatharta, Feardorcha Ó Colla, Timlin Ó Cearnaigh, Seán Ó Loinsigh, Pádraic Ó Concheanainn, Aingeal Ni Chonchúir.**

4.30 DUANTA DURACHTA
ó **Mháirtín Ó Direáin, Máire Mhac an tSaoi, Seán Ó Ríordáin,** is **Séamus Ó Néill.**

5.00 FILLEADH AR AN DUCHAS
Ómós don ócáid ó: **Thomás Mac Diarmuda** (Tom Phaidi), **Pádraic Ó Finneadha, Máirtín Ó Loideáin, Seán Bán Mac Ghrianna, Diarmuid Ó Drisceoil, Nioclas Tóibín, Pádraic Mac Donncha** agus **Proinsias Mac Aonghusa.**

5.30 NA CEOLTOIRI OGA
Buíonta ceoltóirí den aos óg sna dúichí Gaeltachta.

6.00 NUAIOCHT

6.10 TIOMCHUAIRT NA nGAELTACHTAI
Cuid a Dó.
Clár irise, faisnéise, ceoil is reacaireachta a cuireadh i dtoll a chéile sna Gaeltachtai.
Clár-Reachtairi Radio na Gaeltachta a d'eagruigh é.

6.45 AN CHEAD DAMHSA EILE —AMHRAN!
Clár de cheol Gaelach is d'amhráin tíre, de ghreann is de phléaráca dá chur i láthair ag **Seán Ó Tuairisc, Timlin Ó Cearnaigh, Micheál Ó Sé** is **Aingeal Ni Chonchuir.**

7.45 NA CAIRDI CEILTEACHA
Ceol is amhráin is beannachtai ó Albain, ón mBreatain Bhig is ón mBriotáin.

8.15 DEIREADH AN LAE
Diospóireacht ag muintir na Gaeltachta faoi Radio na Gaeltachta.
Peadar Mac an Iomaire dá stiúrú

8.45 NUAIOCHT IS SPOIRT
Scuireadh 9.00.

Luan Cásca

6.00 NUAIOCHT IS TUAR AIMSIRE
Tuar aimsire speisialta do chósta ar Iarthair.
CEOL NAISIÚNTA, AMHRAIN IS FILIOCHT
Clár do Luan Cásca dá chur i láthair ag **Máirtín Ó Fatharta.**
CUIMHNI CÁSCA
Cuimhní ar Cháisc na bliana 1916 ag daoine sna ceantair Ghaeltachta. **Seán Ó Loinsigh** is **Aingeal Ni Chonchúir** a chuir i dtoll a chéile é.
CUAIRTEÓIRI NA CASCA
Agallamh is ceol ó dhaoine atá ar cuairt i gConamar i geomhair na Cásca. **Seán Ó Tuairisc** ina mbrin.

CEOL ÓN ACHRÉIDH
Scothcheoltóiri ó Oirthear na Gaillimhe is ó na Breac-Ghaeltachtaí dá gcur os ár gcomhair ag **Máirtín Ó Fátharta.**

8.00 NUAIOCHT
is **Spóirt.**

Mairt Chásca

6.00 NUAIOCHT IS TUAR AIMSIRE
CUAIRT NA SCOL
Ceol, scéalta is amhráin ó scoltachtaí sa nGaeltacht.
TRATH NA gCEIST
Comórtas idir fhoirne ó na Gaeltachtai dá reachtáil ag **Feardorcha Ó Colla.**
GRÚPAI CEÓIL
Ceol ó bhuíonta sna Gaeltachtai dá sholáthar ag **Timlin Ó Cearnaigh.**

8.00 NUAIOCHT
is **Spóirt.**

Ceadaoin Chásca

6.00 NUAIOCHT IS TUAR AIMSIRE
Seán Ac Dhonncha is a chomlás ceoltóirí is amhránaithe.
MUINTIR NA nOILEAN
Sraith clár faoi na hoileáin Ghaeltachta atá a' síneadh le cósta an Iarthair.
Uimhir 1—**Inis Bearachain is Oileáin Árann.**
CEOL CEIRNINI
Dá gcasadh ag **Aingeal Ni Chonchúir.**

8.00 NUAIOCHT
is **Spóirt.**

Deardaoin Chásca

6.00 NUAIOCHT IS TUAR AIMSIRE
Coimeas idir Cheolta dhúchais na hEorpa dá ríomhadh ag **Micheál Ó Sé.**
Stair an Éirghe Amach. **Gearóld Ó Tuathaigh** dá ríomhadh.
Clár Tighis dá reachtáil ag **Aingeal Ni Chonchúir** agus **Máire Ni Chéide.**
Na Chieftains.
Clár ceirnini dá gcasadh ag **Feardorcha O Colla.**

8.00 NUAIOCHT
is **Spóirt.**

Aoine Chásca

6.00 NUAIOCHT IS TUAR AIMSIRE
Ceirdeannai sa nGaeltacht.
Ríomhadh ar na seancheirdeannai, obair is luadháin is amhráin saothair.
Ceolghrúpai.
Diospóireacht
Faoi chúrsai reatha sa nGaeltacht.

8.00 NUAIOCHT
is **Spóirt.**

After a week they decided that it was more pragmatic to move to Carraroe, 52 kilometres away, where they broadcast for another couple of nights. They thought more people could hear the station from there, perhaps even in the Aran Islands. And listeners did. The pirate station got a letter from a listener in Abbeyfeale, over 200 kilometres away on the border of Limerick and Kerry, declaring that they thoroughly enjoyed listening to Saor Raidió Chonamara. The station broadcast again for a week in November 1970, this time from Tigh Mhico, a well-known local pub in Rosmuc, during An tOireachtas, the annual Irish-language cultural festival. By the end of the month the government announced that RTÉ was to build radio studios in Gaoth Dobhair in Co. Donegal, in Corca Dhuibhne in Co. Kerry and in Connemara, Co. Galway. Saor Raidió Chonamara only broadcast for nine days, but it was nine days that would change the face of broadcasting in Ireland. The act of defiance is recognised as having brought about the opening, on Easter Sunday 1972, of Raidió na Gaeltachta. While its arrival did not officially reduce the requirement nor demand for Irish to be part of the regular offering of RTÉ Radio generally, it was a decisive recognition of an underserved constituency of radio listeners. RnaG went on to become an essential component of the wider RTÉ offering, and its programmes across a range of topics and themes features throughout the book.

The opening-day schedule of RnaG on 31 March 1972.

In 2018, the close-knit community nature of the station was illustrated when playwright and Oscar winner Martin McDonagh

Singer Sandie Jones represented Ireland in the 1972 Eurovision Song Contest with the song in Irish called '*Ceol an ghrá*' written by Liam Mac Uistin and Joe Burkett.

Although RnaG had opened a few weeks previously, RTÉ Radio's Irish-language magazine programme *Idir Shúgradh agus Dáiríre* presented by Diarmaid Ó Muirithe, seen here leaning on the ladder, was the cover of the *RTÉ Guide* on 9 June 1972.

wrote a *míle buíochas* letter in Irish to RnaG's Máirtín Tom Sheáinín request programme *Togha agus Rogha* (The Absolute Best). Mc Donagh wanted to thank those who erected three billboards at Leitir Mealláin in recognition of the BAFTAs he won for his film *Three Billboards Outside Ebbing, Missouri*. A few weeks before, McDonagh's father, Johnny Mhichael Choilm Rua Mac Donncha, who is from Leitir Mealláin in Connemara, had spoken about the excitement around his son's Oscar nomination with RnaG's Máirtín Tom Sheáinín.

Oireachtas na Gaeilge is Ireland's foremost annual Irish-language cultural festival, founded in 1897 by Conradh na Gaeilge (the Gaelic League). A fond, if indirect association I have with the Oireachtas is from schooldays. I had a minor but pivotal part in a dramatisation of the story by Patrick Pearse called *Brighid na nAmhrán* (Brighid of the Songs), a contribution by our school to the centenary of Pearse's birth. Brighid, now an old woman, is determined to perform in a singing competition in the festival. My part had me standing up on a chair as a sort of revolving signpost. In quick succession I displayed signs of several places big and small, from Connemara and Dublin, announcing them aloud and the distance from them to Dublin. The girl in my class who was Brighid walking back and across in front of me, looking up forlornly as the place names changed. My signposting was to illustrate the long, arduous road she took on foot, before she fatally exhausted herself.

RTÉ's *Raidió Scoile* pupils in 1975 were asked to send in something about what they heard on the programme. This is a drawing of the Spanish Armada.

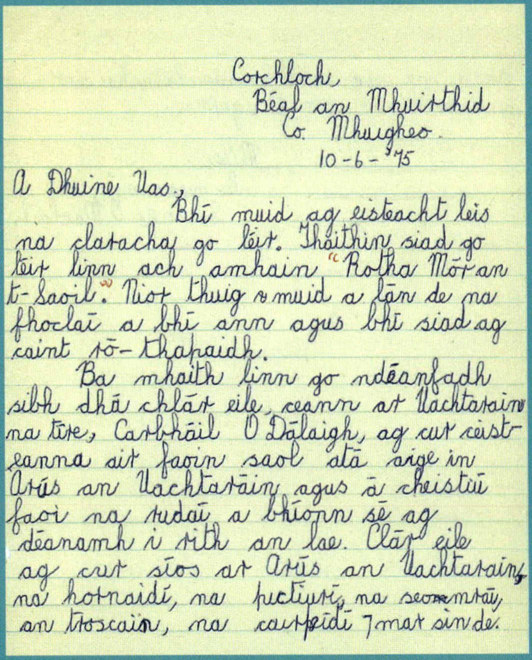

Corchloch,
Béal an Mhuirthid
Co. Mhuigheo.
10-6-'75

A Dhuine Uasal,
Bhí muid ag éisteacht leis na clárachaí go léir. Thaithin siad go léir linn ach amháin "Rotha Mór an t-Saoil". Níor thuig muid a lán de na focail a bhí ann agus bhí siad ag caint ró-thapaidh.
Ba mhaith linn go ndéanfadh sibh dhá chlár eile, ceann ar Uachtarán na tíre, Cearbhaill Ó Dálaigh, ag cur ceisteanna air faoin saol atá aige in Árus an Uachtaráin agus ag cheistiú faoi na rudaí a bhíonn sé ag déanamh i rith an lae. Clár eile ag cur síos ar Árus an Uachtaráin, na hornaidí, na pictiúirí, na seomraí, an troscáin, na cuirtíní 7 mar sin de.

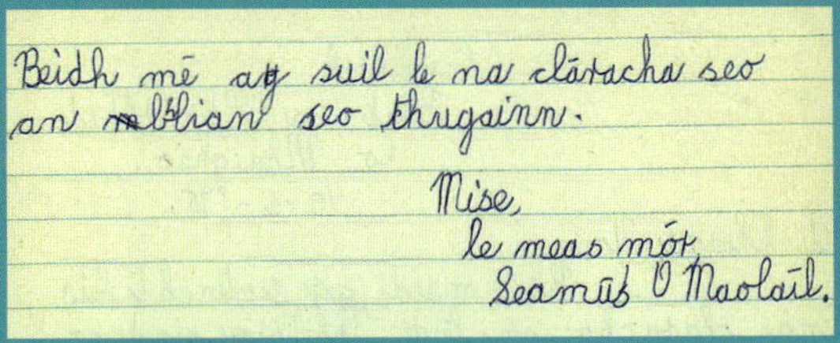

Beidh mé ag súil le na clárachaí seo an bliain seo chugainn.
Mise,
le meas mór,
Seamus Ó Maoláil.

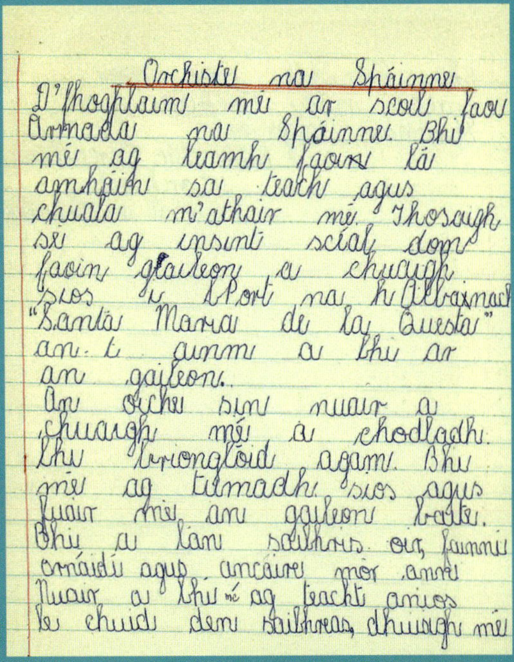

Orchiste na Spáinne
D'fhoghlaim mé ar scoil faoi Armada na Spáinne. Bhí mé ag léamh faoi lá amháin sa teach agus chuala m'athair mé. Thosaigh sé ag insint scéal dom faoin gaileon a chuaigh síos i bPort na hAlbainach "Santa Maria de la Questa" an t-ainm a bhí ar an gaileon.
An oíche sin nuair a chuaigh mé a chodladh bhí brionglóid agam. Bhí mé ag tumadh síos agus fuair mé an gaileon briste. Bhí a lán saibhris ann, fáinní órnáidí agus ancaire mór ann. Nuair a bhí mé ag teacht aníos le chuid den saibhreas, dhúisigh mé

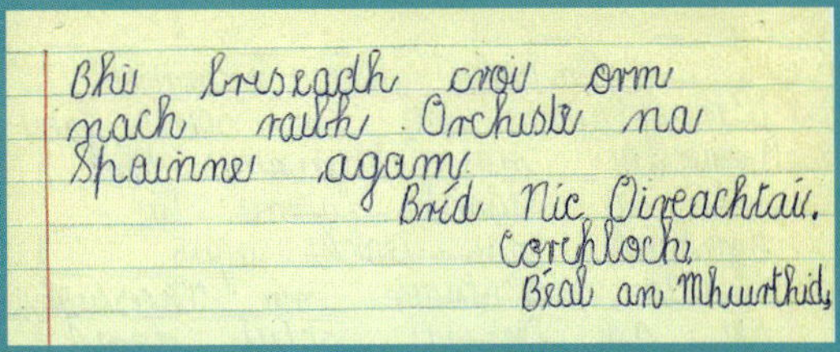

Bhí briseadh croí orm nach raibh Orchiste na Spáinne agam.
Bríd Nic Oireachtaigh.
Corchloch.
Béal an Mhuirthid.

RTÉ's *Raidió Scoile* was an education programme with its own wavelength in the 1970s. When pupils were asked to suggest programmes they would like to hear on radio, one suggested two: President Hillery about life in Áras an Uachtaráin and another describing the building.

Accompanying her drawing the pupil wrote describing her dream of swimming down to one of the Spanish Armada shipwrecks that were said to have had gold in them. Just as she reached the surface of the water she woke up and was broken-hearted that she did not have the gold after all.

RTE GUIDE
PROGRAMMES MAY 22–28 :: Vol. 13 No. 21 :: IRIS RADIO TELEFIS EIREANN, MAY 21, 1976. PRICE 6p.
MEET MAIRE DE BARRA OF ROGHA CEOIL
RADIO: WEDNESDAY

Leagan Cainte was a long-running series about Irish expressions and idioms on RnaG with its own cult following and is enjoyed still today as a podcast. Chaired by Liam Mac Con Iomaire, with panel Ebhlín Ní Mhurchú, Dónall Ó Baill and Seosamh Ó Cuaig.

(Opposite page) Máire de Barra, from West Kerry, was a bilingual radio broadcaster before she moved to television to become a children's presenter and continuity announcer. This *RTÉ Guide* of 21 May 1976 was publicising her requests programme, *Rogha Ceoil*.

In 1974 An tOireachtas decided it was time to hold its festival in a Gaeltacht. RTÉ Radio's programming reflected this when producer Padráic Dolan reported in the *RTÉ Guide* that regular weekly Irish-language programmes would broadcast from the week-long festival. The programmes included news, feature and magazine programmes *Gael Iris, Trácht* and *An Mhuintir s'Againne*, with the intention to provide the spirit and atmosphere of the Oireachtas on the radio. Chiming with the coastal location of the festival, an hour-long programme called *Curracha agus Currachóirí* (Currachs and Currach Men) was presented by Liam Ó Muirthile. RnaG also broadcast at least a daily half hour of the festival's events. RnaG has been covering Oireachtas events live since 1973 and in more recent years events such as *Corn Uí Riada* (Ó Riada Cup) and other *sean-nós* (slow traditional singing in Irish) competitions have proved popular with online listeners globally.

The year 1976 was the fiftieth anniversary of the beginning of 2RN. To salute the occasion, Proinsias Ó Conluain produced a radio programme called *An Ghaeilge agus an Ghaeltacht* for RTÉ Radio. It featured key figures whose voices on the air, as well as behind it, worked with RTÉ Radio in Irish during that time. Ó Conluain mentioned Séamus Clandillon, Conall Cearnach, Máiréad Ní Ghráda, Máire Ní Scolaí, Proinsias Ó Ceallaigh, Seán Ó Riada, Roibéard Ó Faracháin and Aindrias Ó Muimhneacháin,

RTÉ Raidió na Gaeltachta's *Tús Áite* broadcasting live from the GPO on 15 January 2026 as part of RTÉ100 celebrations. From left, Sighle Ní Bhaoill, retired sound engineer, presenters Doireann Ní Bhriain and Fachtna Ó Drisceoil, author Iarla Watson and former RTÉ archivist and programme maker Ian Lee.

who progressed and expanded broadcasting in Irish. Seán Mac Réamoinn, who in 1976 was controller of RTÉ Radio programmes, thanked the many people who spoke on RTÉ Radio in Irish and poignantly remarked that, though they had passed on, the broadcasts made of them, and now archived kept their Irish speaking voices alive. He said that the story of Irish-language broadcasting had multiple facets: many to be proud of and many to the contrary.

In 2019 popular Irish-language presenter and DJ Seán Bán Breathnach – or 'SBB' as he is also known – celebrated fifty years on air. RnaG's weekday news and features programme, *Tús Áite,* broadcast a programme to mark the occasion. Breathnach did his first live broadcast on Radio Éireann when still a teenager, though he was already an experienced DJ on pirate stations and in clubs in England part-time, while working on building sites. His first Radio Éireann programme was *Pop Seó na Máirte* (Tuesday's Pop Show) produced by John Keogh, formerly of Dublin music group The Greenbeats. *Pop Seó na Máirte* was a trailblazer, being the first Irish-language pop show and the first programme on Radio Éireann in which a DJ could ad lib. Breathnach played sets of reggae, soul and Motown. Some listeners told SBB he would destroy the Irish language, while others loved what he was doing, even requesting a lock of his golden curls!

Breathnach started in RnaG in 1972, shortly after the station went on air. His passion is sport, and he has anchored *Spórt an tSathairn* (Saturday Sport) on RnaG since 1978. Breathnach was inducted into the PPI Broadcasting Hall of Fame in 2012, the same year his commentary in Irish on Katie Taylor's gold medal win at the London Olympics went viral. The commentary was included in listeners' sports commentary moments during the RTÉ marathon cross-station broadcast from the GPO as part of RTÉ100 in January 2026.

Other standout Breathnach broadcasts are his commentary on the boxing match of Seán Ó Mainín in Madison Square Garden, Galway's All-Ireland Senior Hurling Championship win in 1980 and his interview with the Hothouse Flowers in 1988 when they performed 'Don't Go' on BBC's *Top of the Pops*; Breathnach recalled the great respect the group showed for RnaG.

In 2022 Irish was fully designated an official language of the European Union. The process began in 2007, when Irish was recognised as a working language. The language's status meant that advertising in Irish and general public information announcements are integral sounds of RTÉ's radio stations; and the same announcements in both English and Irish mean listeners are more likely to grasp the one in Irish after a few broadcasts.

Beo ar Éigean (Hardly Alive) began as a podcast in 2017 before having a Saturday evening Radio 1 slot. It captures the honesty and humour of three good friends as they navigate modern life. The presenters are RTÉ online Gaeilge editor Siún Ní Dhuinn, Sinéad Ní Uallacháin, whose voice is known to so many as one of RTÉ Radio's anchor announcers in Irish and in English, and Áine Ní Bhreasláin of RnaG.

This chapter could not be written without the inclusion and acknowledgement of Manchán Magan and the sea-change of positive engagement with the Irish language, the natural and ancient world which he brought to another generation of radio

RTÉ

(and television) listeners. Like Seán Mac Réamoinn said about an earlier generation of broadcasters, it is a gift to listeners that Magan's voice is now preserved in RTÉ Archives.

Since its establishment in 1972, RnaG has expanded a full broadcasting service for the Gaeltachtaí and for Irish speakers all over the world. Nuacht and weather are broadcast daily on Radio 1. There are occasional feature series, plays and documentaries. Contributions in Irish feature in RTÉ radio-writing competitions such as the RTÉ Short Story Competition in memory of Francis MacManus, and commissioned and open series like *Sunday Miscellany* and *Spoken Stories*. Evelyn O'Rourke, and before her, RTÉ's first designated Arts and Media Correspondent, Sinéad Crowley, reports in English and as Gaeilge.. Artists and their work in Irish feature on Radio 1's *Arena*. There is a deep dive into all things Gaeilge during Ireland's annual *Seachtain na Gaeilge*. Programme presenters, of their own volition, sign off with *slán go fóill* (bye for now). Evolving technologies mean broadcast programmes in English and Irish are available on demand, something unimaginable for so much of RTÉ's first 100 years. Whether the measure of these meets the broadcaster's statutory obligation to Irish or what listeners would like to hear is an open question.

Manchán Magan in 2021.

(Opposite page)
Top: Public participant enjoying RnaG's live broadcast of *Bladhaire* from The Piper's Corner in Dublin at Christmas 2019.

Middle: 2FM's Hugh Mac Giolla Chearra with RnaG's Áine Ní Bhreasláin at same event.

Bottom: Radio 1's and podcast editions of *Beo ar Éigean* presented by Siún Ní Dhuinn, Áine Ní Bhreasláin and Sinéad Ní Uallacháin featured them talking their life and times with its own mix of humour and poignancy.

VISUALLY SPEAKING

Jack B. Yeats

In a brief Radio Éireann recording from 1947, Jack B. Yeats told broadcaster Éamonn Andrews that he was not at all fond of talking about his own or other painters' work. He explained that he never lectured on painting and had never even asked a question at a lecture, though he greatly enjoyed listening to others speak. Perhaps, then, Yeats would have enjoyed hearing the director of the National Gallery of Ireland, Thomas Bodkin, when he gave a radio talk on 2RN in 1928. He may have also listened to the 1935 discussion series that featured artist Mainie Jellett on 'the Academic tradition and the Modern Art Movement'. Or perhaps he tuned into those broadcast in the 1940s on modern painting featuring artists such as Seán Keating, Louis le Broquy and Tom Collins.

Indeed, it is reasonable to wonder what Yeats would have made of broadcaster Marion King. Her Radio Éireann programme *Drawing and Painting* began in the early 1940s and ran for decades; it was popular with young and adult listeners alike. Children of all ages sent her their art, inspired by the character Sean Bunny, Cushaboo and their friends, which she created. She also organised annual exhibitions of art originating from the programme. King was the trailblazer for children's radio programmes that encouraged visual art and crafts activities, well before the arrival of children's television programmes, at a time when art was an ancillary subject in school, a reward after other lessons, often relegated to Friday afternoons.

Though writer Monk Gibbon had presented mid-century arts programming, a weekly programme became a staple from the early 1970s. *Appraisal* presented by Des Hickey, for example, covered events like the Cork Film Festival and the Dublin Theatre Festival and reviewed plays and dance shows, films and books, as well as interviewing artists associated with them.

COMING BROADCASTS

Art, Architecture, and Literature

Leading Irish men and women in Art, Architecture, and Literature will participate in three new features arranged for by the Saorstat Broadcasting Station.

One is a series of discussions by prominent Irish artists on various phases of art in this country. The first talk in this series will be given on October 1 by Mr. Dermod O'Brien, President of the Royal Hibernian Academy, who will debate with Miss Mainie Jellett on "The Academic Tradition and the Modern Movement in Painting." Mr. George Atkinson and Mr. Sean Keating and other artists will take part.

On October 7, Mr. H. Allberry, the President of the Royal Institute of the Architects of Ireland will open a group of six talks in which Architecture will be discussed. The others taking part are all well-known Irish Architects, Prof. R. M. Butler, Mr. J. J. Robinson, Mr. Vincent Kelly, Mr. Manning Robertson, Mr. A. Pugin Meldon, and Mr. Henry H. Hill, Cork.

Mr. T. C. Murray, the well-known author, will open on October 9 a series of readings by authors of unpublished short stories. Prof. Daniel Corkery, Mr. Frank O'Connor, Mr. Sean O Faolain, Mr. Francis Stuart, and other Irish writers will take part.

Clockwise from top left:
The *Evening Herald* in 1935 mentions a series of talks on art and architecture to be broadcast on the 'Saorstát Broadcasting Station'.

Lorna Madigan, presenter of *Club na nÓg* programme for young people in 1976, sorting out pictures, masks and models sent in by young listeners.

Sculptor Cliodna Cussen with producer Pat Feeley recorded for the 1978 radio portrait *Stonies*.

Errand boy in King Street, Dublin, in 1952. Photographer Nevill Johnson talked to Freda McGough on *Appraisal* in 1988 about his new exhibition at the Gallery of Photography in Dublin.

Marion King recording her art programme with characters Cushaboo and Sean Bunny in the GPO studios in the 1940s.

***The Arts Show* presenter Mike Murphy and producer Maura Lee examine a painting by Brian Bourke in the Royal Hospital, Kilmainham, in 1989.**

Aedín Gormley in conversation with Angela Lansbury in 2016 at the Bord Gáis Energy Theatre for her lyric fm *Movies and Musicals* show as part of the Dublin International Film Festival.

However, the launch of *The Arts Show* broadcast three days a week with Mike Murphy at the helm in 1988 augured a new commitment by RTÉ to the arts. Not that everyone was convinced that Murphy was the right person for the job – some thought he was too populist and breezy a personality for the serious (and elite) manner in which the arts should be treated. Listeners proved the show's popularity, and it began to be broadcast daily throughout the week, directly after *Liveline*. Murphy's obvious passion for visual arts led artists to trust him, though he often asked provocative questions, sometimes baffling them, but in the process getting them to talk, and keeping listeners with the programme. Looking back, Mike Murphy's time on *The Arts Show* was a daily mini arts festival on the air.

When RTÉ's music and arts station lyric fm was launched in 1999 its arts programme was *Artzone* presented by Aedín Gormley. Her visual slant won out for her as she went on to forge an identity on the station with cinema and musical theatre. Luke Clancy, who had been a visual art reviewer for *The Arts Show*, established a regular programme called *Culture File* on lyric fm that continues still. For The Lyric Feature on lyric fm writer and art critic Cristín Leach brought listeners a series of programmes she called *Ireland Portrayed*, an exploration of how and what was revealed visually through artworks about Ireland and its people over three hundred years.

Kay Sheehy, Ireland's first county arts officer and a long-time producer at RTÉ, understood the importance of engaging public interest in the arts via emotionally impactful stories. This was the thrust of *Eye Candy*, her early-2000s series.

A little over a decade earlier, in 1993, Brian Kennedy, assistant director of the National Gallery of Ireland told Bryan Dobson on Radio 1's *News at One* about the authentication process of Caravaggio's *The Taking of Christ*. It had been housed in the Dublin Jesuit House for decades. Following its restoration, it was to be on permanent loan to the National Gallery of Ireland. It was later featured on Sheehy's programme, *Eye Candy*, in which were interviewed the gallery's senior curator, Sergio Benedetti, and Frank McGuinness, whose play *Innocence* followed a day of the life of Caravaggio. When I heard her programme on Velázquez, it brought me back to wintertime Madrid, when I lived there, and to queuing for hours to see his art in the Prado, and to the mixture of familiarity and pride I felt on seeing his painting *Kitchen Maid with the Supper at Emmaus* and the label beside it that declared it 'on loan from the National Gallery of Ireland'. Listening to the radio can take us places ...

Diego Velázquez (1599–1660) *Kitchen Maid with the Supper at Emmaus* in the National Gallery of Ireland Collection.

THIS ISLAND NATION

CHAPTER 8

Inis Oirr, looking towards the coastline of Co. Galway.

Reports of the day say the weather was calm. I cannot remember whether I saw the narrow upright thread of smoke rise almost straight up against the blue sky or heard about it on the radio first. It was 16 August 1971, and McDonagh's warehouse in Galway city was on fire, along with several other buildings around it. Our family was on summer holidays at the time on Inis Oirr, the smallest of the Aran Islands. It was the first time I became conscious of clearly looking across to the the rest of Ireland, watching something strange happening within eye view, yet somehow aware also of the safety that comes with distance – over two hours of a journey on the edge of the Atlantic on the *Naomh Éanna*, the vessel that had brought us to the island, separated me – and everyone else on the island – from the raging fire.

It was the same ocean I heard Mary Lavin talk about with John Quinn in 1985 on his radio series *A Portrait of the Artist as a Young Girl*. She told him how, for her, the sea was an emotion, because when she was a girl the Atlantic had separated her from her father. She travelled from America to Ireland for the second time in 1921 at the age of nine on the SS *Winifred* with her mother. They had been brought to the ship by her father, whom she saw drive away in his beach wagon car (he would join them in Ireland a year later). Although she knew she was going to see him again, it was a lonely thing to see him drive away because she and her mother were 'going across the whole Atlantic from him to Ireland', as she put it. When she got to her mother's family in Athenry in Co. Galway and they'd go to the sea in Salthill or wherever, she had no interest in making sandcastles. She could make them when she went to the sea at Bettystown or Killiney on the east coast. Meanwhile the great stretch of water that was the Atlantic did nothing for her but separate her from her father. A different sea altogether. On air, Lavin's story of the sea and distance evoked emotion in the listener as much as it had for her.

Writer Mary Lavin.

While an island is generally defined as land surrounded by water, visible at high tide, it is tricky to account for the exact number of them that dot the coast, many of them along Ireland's western seaboard. At the time of the 1926 Census nearly sixty islands had people living on them, while in 2026 about half that number are inhabited on a year-round basis. Islanders depended on local, diesel-run generators for power until the late 1970s and 1980s, so regular, battery-run transistor radios were common.

The official resettlement of the last twenty-two inhabitants of the Great Blasket Island commenced on 17 November 1953. Over the following year they were evacuated to the mainland to housing provided for them by the State.

The islanders increasingly felt the isolation, lack of medical and emergency supports and severe weather that often cut them off from the rest of the country for long periods – and without the assistance of younger, able-bodied residents. This reflected general migration trends nationwide, where younger people were

The evacuation of the Great Blasket Island, November 1953.

lost to rural areas through emigration and movement to towns and cities for employment. Visitors to the island encouraged the islanders to document their lives, to tell their stories. Out of this came books like *An tOileánach* (The Islander) by Tomás Ó Criomhthain (1929), *Fiche Blian ag Fás* (Twenty Years A-Growing) by Muiris Ó Súilleabháin (1933), *Peig* by Peig Sayers (1936) and *Méiní, the Blasket Island Nurse* by Leslie Matson (1996). Over the years these books were serialised and read on air. The books, the Blaskets and its people inspired many other radio programmes about island life in general.

The Radio Éireann mobile recording unit was established in 1947, with staff including Seán Mac Réamoinn and Séamus Ennis recruited. They made their first foray out of Dublin in 1948, carrying the recording equipment on the back seat of Ennis's car to begin gathering material that could be used for radio programmes on the station. Peig Sayers's voice was among the

Tomás Ó Criomhthain, author of *An tOileánach* (The Islander) in his home on the Great Blasket Island read for RTÉ Radio's *Booktime* by Niall Tóibín and first broadcast in 1978.

▶ Peig Sayers while she was still living on the Great Blasket Island.

first they recorded. By this time, Sayers, the Kerry storyteller, was back living in Dún Chaoin, where she had started her life before moving to the Great Blasket Island. Like many others she had hoped to emigrate to America, but when money for the journey failed to materialise, she had no option but to continue in service until her father rescued her from a harsh employer by arranging for her to marry islander Pádraig Ó Gaoithín.

Mac Réamoinn related how 'news came that Peig ... was dying, or was very ill, and that it might be possible to record something from her before she died, it sounds a bit ghoulish when you say it like that ... and Peig I'm glad to say, lived for many years after that. But our visit to Peig that night is in itself a saga, and in fact Peig has told the story herself':

> Ach ní a fhad gur tháinig an fothrom agus an solas go dtí an doras, gur phreabas, dá laigeacht a bhíos agus dá mhéid greann a bhí orm, ach gur shiúl romham isteach triúr strapéirí fear agus d'fháiltíos rompu de réir mar ba cheart dom a dhéanamh. 'Ach cad a thug sibh?' arsa mise leo. 'Do thug ...' arsa síodsan, iad ag tarraingt boscaí agus giúrléidí agus rudaí trína chéile isteach go raibh an tinteán lán de ghiúrléidí thíos, radharc nár facthas ariamh agus ná chifead go deo arís faid a bheidh seana-Pheig mar ainm orm. Ach cad a bhí do bhí – rud éigin a bhaineas le radio nó rud éigint acu nárbh fhios domsa.

> It wasn't long before the noise and light came to the door, that I jumped, delicate as I was, and felt such surprise when three fine men walked in and I welcomed them in as I should do. And what brought you? I asked them. It was, they said, something to do with radio or something that I didn't know, as they pulled in boxes and equipment and a mix of all sorts of things, until the fireside was full of them, such a spread of things as I had never seen and would never see again as long as my name is old Peig.

Cover of *Valentia: Portrait of an Island* by Daphne D.C. Pochin Mould (Blackwater Press 1978).

For his 1981 series *Authors and Islands*, producer Proinsias Ó Conluain asked writers to talk about books about islands around the country. Nóra Ní Shúilleabháin chose *Valentia: Portrait of an Island* by Daphne D. C. Pochin Mould (1978). The first successful transatlantic telegraph cable was laid between Valentia and Newfoundland in 1866, connecting North America and Europe and changing world communications for ever. And although Valentia had been referred to in books on subjects as various as geology, meteorology, botany, marine biology, telegraphy, wireless and folklore, Pochin Mould's was at the time the only book specifically about the island and the lives of those who called it home. Ní Shúilleabháin declared that she had better 'set in the island in its place for listeners' and on she went with the details, bringing us there in the process:

> It is in the barony of Iveragh, southwest of Cahersiveen, seven miles long by a maximum three miles wide. It stays within the outline of the mainland on two sides sheltered by its hills. Most of the northwest side is exposed to the north winds that come coldly down from the north while the west faces the immensity of the Atlantic Ocean. Beyond the continental shelf the ocean floor sinks to ninety fathoms and over and in from that cauldron come the warm southwest anti-trade winds. These prevailing winds produce an equitable climate, no extremes of cold but super rainstorms that break their initial fury on the Iveragh headlands. The wild sea has taken its toll and there have been several losses of fishermen. Riders to the sea again and again ...

Mary O'Shea, Valentia local, gave accounts of island life to Pochin Mould, who noted O'Shea was gifted with an acute sense of the comical. She had a song about a rat she commanded to be gone and to go instead to all the other houses on the island. The rat must have known Irish, because he took her directions before dying in the last house named in the song, old and small as a mouse killed by a fire lit by the woman of the house. Nóra Ní

Shúilleabháin concluded her contribution on the book about Island with the observation that 'in a sense no island is an island anymore ...' For an island so central to early modern communications, those words of over forty years ago ring true, at least as far as communication goes.

Provincial Round-up was a weekly programme introduced in the 1950s to feature life outside of Dublin. In the end-of-year programme of 1959, presenter Karl Jones told listeners that at that moment a plane was flying over the midlands that had just dropped provisions to the western islands marooned by gale conditions. Supplies contained tea, whiskey, the day's newspapers and all kinds of Christmas fare. The plane flew over Inis Shark (evacuated in the 1960s), Inish Turbot (evacuated in 1978) and Inishturk (with a recorded population of fifty to sixty in 2025) off the Connemara coast. It had to do a practice run over each of the islands before dropping its supplies. Proinsias Ó Conluain made a programme documenting Christmastime on Inishturk in 1974, anticipating that it might follow the fate of other islands left with no inhabitants.

Though considered that it might follow other islands to become uninhabited, today Inishturk continues to have a small, thriving population, here enjoying its annual regatta in 2026.

(Opposite page)
The funeral procession following the sinking of the *Lusitania* in 1915.

On *Highways and Byways*, the 1970s and 1980s weekly programme, presenter Donncha Ó Dúlaing travelled the country meeting people and bringing their stories to listeners. In 1978 he got a first-hand account of the day on 7 May 1915 when the *Lusitania* was struck by a torpedo off the Cork coast during the First World War. Gerry Murphy told Ó Dúlaing that he was sixteen, gathering seaweed that morning when the coxswain of the local lifeboat came running up to him. Together they ran to where the lifeboat lay and with a crew of eleven they rowed to the bodies in the water. It took the men three hours, as there was no wind to assist them. Steadily they began the task of lifting bodies into their boat. Gerry Murphy gave graphic details of the bodies they brought back to land with them, including a Swedish butler in uniform, a stoker, a young girl and old men.

Sadly, there have been many other tragedies at sea that were conveyed on the radio news and later in documentaries or in interviews and conversations.

(Opposite page)
Aerial view of the Whiddy Island disaster on 9 January in 1979.

What happened in the early hours of 8 January 1979 remains the greatest maritime disaster to have occurred in Irish waters since independence. Whiddy Island is only minutes off the coast of west Cork, near the town of Bantry. The island housed Ireland's only oil terminal, owned by Gulf Oil. Whiddy Island drew the attention of the world when the oil tanker *Betelgeuse*, which was owned by international company Total Oil, docked at the offshore jetty and exploded, killing fifty people. Only twenty-seven bodies were recovered. Breandán Ó Cíobháin's 1979 programme *Disaster at Whiddy* captured the first-hand, contemporary experiences of the tragedy. The 2021 documentary *Fire in the Sky*, produced by Donal O'Herlihy and Michael Lawless, focused primarily on the personal story of 31-year-old Whiddy pollution officer Timothy Kingston, who lost his life that night. Key to the impact of the programme is the contribution by one of his three children, Michael, whose fourth birthday it was the day before the tragedy, and who is now a maritime lawyer. Forty-four years later in 2023, Michael Kingston and his mother, Mary, spoke with Miriam O'Callaghan on her Sunday-morning programme. Each of the three radio programmes on Whiddy gave the listener a different perspective on the story, from when it was first introduced as breaking news to a more nuanced, deeper understanding through the documentary and then more personally, emotionally perhaps, through conversation on air.

The hands of the King of Tory taken by Cian McCormack during their walk in Tory for *Morning Ireland* in 2014.

The Wild Atlantic Way opened in 2014 with the intention of encouraging sustainable tourism along the west coast of the country. It linked up existing roads and routes and 4,000 signposts were erected for those travelling along its length. It also provided an excuse for Cian McCormack's summertime bike trip from which he gave live daily reports to *Morning Ireland* that summer. From Tory Island he spoke live to presenter Frank McNulty, beginning as Gaeilge: '*Maidin mhaith ó Oileán Toraigh agus tá fuinneamh sa spéir freisin. Tháinig mé anseo ag meán lae [inné].*' (Good morning from Tory Island. And there's some strength in the air. I arrived in the Ferry at noon [the day before] and every visitor was greeted by the King of Tory, Patsy Dan McGory.) And as he walked along the island with McGory and Brigid McKervey of the island co-op, they shared with listeners realities of twenty-first-century island life. They spoke about the desire for more visitors and how the Wild Atlantic Way could assist; they talked about the isolation of not being served by airplane, of the 140 islanders having use of a helicopter for thirteen annual journeys for planned medical, dentistry and other appointments. They joked about bad hair days without a hairdresser on the island. The key priorities, as far as the king of the island was concerned, were continuing Tory's music and song traditions and maintaining the Irish language.

Happily, short- and long-term visitors have been drawn to the islands. On *The Ryan Tubridy Show* in February 2019, Lesley Kehoe told Tubridy how she and her partner, Gordon, were planning to spend from April until October on the Great Blasket Island as caretakers, looking after the shop and accommodation – quite the while to spend on an island with no electricity or hot water. It would be a culture shock, but one that they were willing to embrace.

> It's probably the darkest place I've been to: no lights, no interference from the outside world ... the nicest thing is sitting back and listening to the seals chat to each other on the beach when it gets dark, you have to forgo the books at that stage.

'Good evening and welcome to *Seascapes*, the maritime programme.' These were the introductory words to RTÉ's first radio programme dedicated to the sea. It was presented from 1989 to 2019 by marine correspondent Tom McSweeney, then by Marcus Connaughton, Fergal Keane and Lorna Siggins. The calming signature tune of 'Sailing', composed by Ronald Binge (the same music as BBC's shipping forecast), is loved by listeners, as was made clear by blowback from audiences when the tune was briefly changed in 2009 and it wasn't long before it was reinstated.

Volunteers and voluntary organisations have been a regular source for material featured on *Seascapes*: the RNLI, Inland Waterways Association of Ireland, the Irish Coastguard, Irish Sea Scouts, the Irish Sailing Association and coastal sub-aqua clubs. Many listeners are themselves members of boating, fishing, rowing and swimming clubs and associations. A highlight for McSweeney was the annual national sail-training awards and interviewing the youngsters who took part.

Master boat-builder Joe Murphy, born in 1928, talked in 1991 about his family's Dublin boatyard and how, after the Second World War, people who had been in the army received gratuities with which to buy boats, so that it was no longer only the wealthy who sought the services of the boatyard. A compulsory purchase order was made of the land that had been in his family for generations because the city council said they were not providing enough employment for the expanse of land their business occupied, and Murphy went to work at the Dublin Port and Docks Board. Eight blocks of flats for Dublin Corporation were built on the land his grandparents had acquired over time. As he talked, looking out to sea, Murphy noticed a Galway hooker, the *Morning Star*, and said he would like to get involved in building one of them, and that he would build boats for as long as he lived.

The *Naomh Éanna*, on which I had voyaged to Inis Oirr, was the subject of a report on *Seascapes* in 1999 when it was returned to its first dry dock in Dublin's Ringsend. Plans were in train to turn it into a museum. Prior to the introduction of its light airplane service, the only regular way to travel between the Aran Islands and Galway was by the *Naomh Éanna*, operated as the maritime division of CIÉ. Its launch had been reported by Radio Éireann's *City Reel* programme forty years previously. On *Seascapes* in 1999, former crew members reminded listeners that only the worst winter weather prevented her from travelling (as well as a brief strike in the early 1970s). She could carry up to 320 passengers, and was often full of teenagers travelling to summer college on the islands during the warmer months. She could also carry up to fourteen cows at the same time as general cargo, and an occasional car. The introduction of a high-speed ferry from Rossaveal was due to begin when the broadcast was going out in 1999. Neither listeners nor those reporting the story could have known the future fate of the *Naomh Éanna*.

The *Naomh Éanna* beginning to show signs of neglect, docked in the Grand Canal Dock, in Dublin's Ringsend, not far from where it was launched in 1959. It was sold for scrap in 2024.

The land, and agriculture specifically, was a key focus of 2RN in the early days. *Honesty* magazine on 19 November 1927 reflected an early understanding of Irish broadcasting as a device to inform the public. The magazine reported that a series of talks for farmers was to be broadcast by Mr McGovern of the College of Science. The undisclosed reporter wrote that they had suggested as much to the station's administration and that they were glad to see that the talks were going ahead – agricultural education was of paramount importance in Ireland – and that the radio was the finest medium for its development.

Station director Séamus Clandillon, with a mind on listeners more generally, wanted talks to be engaging programming, while the focus of the department of agriculture was on getting information out by experts to farmers, no matter how it was delivered. A long-standing tension between the two views would rumble. By the time another station director, T. J. Kiernan, was in place, the balance had swung in favour of a more authentic voice representing agriculture on the radio, as John G. Litton, a farmer in Kilternan in Co. Dublin presented his own series, *A Farmer Talks Shop,* in 1945 and 1946.

At that time Radio Éireann was still operating directly under a government department. So the suggestion by minister for agriculture, James Dillon, in 1956 that a member of Radio Éireann's staff would be made available to liaise directly with his department on agricultural programming shoudl not be construed as it might be if it were being asked today. Desmond Fricker, who did many reports relating to agriculture for the station, was seconded to the department from Radio Éireann. The significance of the radio, alongside the newspapers, to connect with the public on matters of national interest and agriculture had been harnessed before and would be again.

(Opposite page)
Many companies did release workers to travel to give support to gathering the harvest of 1946, including many from Jacobs Biscuit factory.

Radio has an immediacy that allows it to not only spread the word but galvanise and involve the country. As taoiseach, de Valera had been a regular speaker on Radio Éireann over the years of the Second World War. By 1946 the war was over, but a domestic crisis struck. The wheat harvest that year was in imminent danger of being destroyed because of ferocious rainfalls. Such a situation would have been serious in any circumstances, but it was doubly so because the world's shipping capacity had been significantly reduced during the war so we couldn't simply import grain from elsewhere. Ireland was still under heavy food rationing. Neither was continental Europe able to assist, as it was only getting to grips with its own post-war rebuilding. De Valera gave a broadcast on 21 September about the harvest situation:

> What is at issue is our food supply ... and the whole community owes a debt of gratitude to all who have assisted – to the farmers and their employees who have worked steadfastly under the most adverse conditions, to the many volunteers who are leaving their shops and offices to serve in the fields at unaccustomed work; to the business firms and public bodies who have released their staffs, and to those who are providing much needed transport; to members of the Irish Red Cross Society and to those who are helping them by providing gifts of food or subscriptions for food; to the military and Garda who are giving help, invaluable help, in many forms to the press, radio and the local committees without whose co-operation the organisation which the crises brought forth could not have been successfully established ... We are grateful to the many sports and recreational bodies who have postponed, at considerable inconvenience and loss, their important weekend fixtures ... It is the duty of everyone who can do so, even at some sacrifice, to assist ... No one who can at all help should hold back. The weekend will provide an opportunity for those who have been unable to get free at other times.

(Opposite page)
The broadcast of An Taoiseach Éamon de Valera makes the headlines in the newspapers, including the *Irish Independent* of 21 September 1946.

The *Irish Independent* reported that de Valera spoke in English and in Irish, calling for an 'all-out' drive to save the country's wheat harvest. Two days later, Minister for Agriculture James Ryan broadcast for twenty minutes on the harvest situation to address farmers and thank the people for coming together. *The Irish Times* led with his broadcast the next day.

CALL FOR "ALL-OUT" HARVEST DRIVE

Mr. de Valera Fears Bread Rationing

Larry Gogan

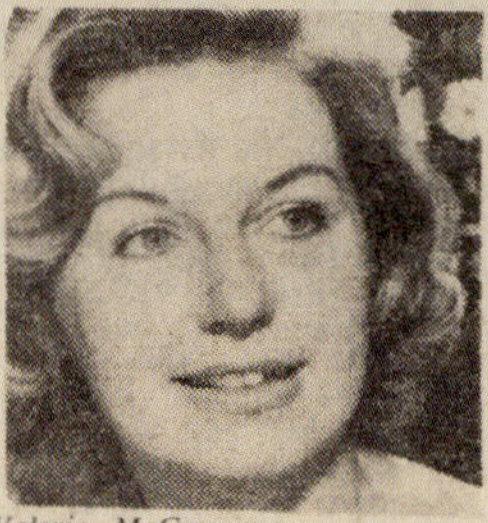

Valerie McGovern

Ciarán Mac Mathúna

JOIN US AT THE SHOW

MAY-TIME again, Spring Show time again . . . once more the famous Ball's Bridge grounds of the Royal Dublin Society will be alive with the bustle of visitors to the traditional Spring Show. RTE's team will be there, as usual, to meet visitors and bring favourite programmes from the now familiar RTE Stand that looks out on the judging rings from the back of the Main Hall. And whether they are at the Show or not, farming people, will be interested in the three special agricultural programmes, produced by Michael Miley.

TUESDAY, MAY 1

10.01-10.58: The Gay Byrne Hour at the Spring Show.

12-15-12.50: Spring Show Requests from the Spring Show, presented by Larry Gogan.

4.01: Spring Show Agricultural Report presented by Michael Dillon.

4.31: Hospitals' Requests from the Spring Show presented by Aidan O'Hara.

5.30-6.30: Music On The Move from the Spring Show presented by Valerie McGovern.

WEDNESDAY, MAY 2

10.01-10.58: The Gay Byrne Hour at the Spring Show.

12.15-12.59: Spring Show Requests presented by Pat Kenny.

4.31: Agricultural Magazine from the Spring Show presented by Larry Sheedy.

5-30-6.30: Music On The Move from the Spring Show presented by Valerie McGovern.

THURSDAY, MAY 3

10.01-10.58: The Gay Byrne Hour at the Spring Show.

12.15-12.59: Ciaran Mac Mathuna at the Spring Show.

5.30-6.30: Music On The Move from the Spring Show presented by Valerie McGovern.

FRIDAY, MAY 4

10.01-10.58: The Gay Byrne Hour at the Spring Show.

12-15-12.59: Spring Show Requests presented by Treasa Davison.

4.01: Spring Show Agricultural Round-up presented by Matthew Dempsey.

5.30-6.30: Music On The Move from the Spring Show presented by Valerie McGovern.

The RDS Annual Spring Show was a popular attraction from the 1970s until it ended in 1992.

From the late 1960s, radio was forced to reckon with the fact that television was cannibalising its audience and to consider where its own strengths lay and what it could do differently to keep farmers listening. *Country Call* started broadcasting early on Sunday evenings and focused on agricultural current affairs. During the same period, *Farming Progress* provided a forum for debate, with broadcasts from rural areas and farmers taking part in on-air discussions. Meanwhile, *On the Farm* was more instructive about changing technical aspects of farming, but it also included some colourful elements. For example, while *On the Farm* may have opened with the serious business of the price of blood stocks, one episode followed with the reaction of city children on visiting a farm. A young girl declared that a field 'was all lovely ... all grass, like a green blanket. It was gorgeous and there were loads of daisies ... I'd say it's about two miles. It's very long and wide, at the side of it there's gates for the cows and bulls to go in.' A boy looking in at the hens who has been told they are about to lay eggs added: 'I'll cook it for my breakfast ... where does it come out? ... does it come under there? ... oh, look at the back of it ... it's moving! ... oh it's a beauty ... You have a baby now.' The facts of life on a farm ...

By the 1970s *On the Farm* included a regular contribution from Brussels to keep listeners in touch with developments in the EEC, as the country was experiencing its first decade as a member.

(Opposite page)
The *RTÉ Guide* featured the first European elections, a reflection of the interest being taken by the population in general since Ireland's recent membership of the EEC, especially so in rural Ireland and in farming.

RTE GUIDE

Vol. 3. No. 22. JUNE 1, 1979. Programmes June 2 — 8. Price 15p (inc. V.A.T.)

IRELAND VOTES on Thursday

First European Parliament Elections

SEE INSIDE

Countrywide also offered news and feature stories related to the land when it started in 2009. Its first presenter was Damien O'Reilly, whose agricultural news reports on Radio 1's *Drivetime* programme were already popular. Other presenters of *CountryWide* have included Ella McSweeney, Marty Morrissey and, more recently, Philip Boucher-Hayes. Each of them – while focusing on their own areas of knowledge from biodiversity to the changing nature of rural communities to more urgent campaigns on the health and protection of our natural resources – continued the long tradition of informative and entertaining broadcasting on rural life in Ireland.

Radio 1's *Countrywide* broadcast a Christmas edition in 2017 from Kylemore Abbey, a Benedictine monastery founded in 1920 on the grounds of Kylemore Castle, in Connemara, Co. Galway.

***Countrywide* broadcast live from the cattle shed of then Irish Farmers Association president Francie Gorman in Co. Laois during the fodder crises in 2018.**

A working Irish farmyard (date unknown).

Farmers protesting in 1978.

Irish Farmers' Association leader Pádraig Walshe (left) and chairman of Farmers for No James Reynolds (centre) talk to broadcaster Pat Kenny in Arnotts department store on Henry Street, Dublin city, during the live broadcast of *Today with Pat Kenny* on Radio 1 on 29 September 2009. This edition of the show featured a debate on the Lisbon Treaty.

I wonder what made Alice Taylor's memoir *To School Through the Fields* an overnight bestseller on its publication in 1988. It was an account of her rural Cork childhood in the 1930s and 40s. She spoke with Gay Byrne on his radio show on the eve of its publication. The interview illustrated the power of the programme, with its hundreds of thousands of listeners. The next day the first printing of the book sold out. The publisher, Brandon Books, immediately reprinted 15,000 copies, which again vanished in days. *To School through The Fields* was also recorded for Radio 1's *Book at Bedtime*.

Alice Taylor, author of *To School Through the Fields*.

Many rural listeners must have recognised their own lives in Taylor's book. Its publication came at a time of aggressively fought referendums on personal freedoms from divorce to abortion. Shocking, dark stories had recently played out in rural Ireland: among them the death of Ann Lovett and her baby in Longford and the accusations against Joanne Hayes that she had murdered her own child in rural Co. Kerry. Here was a reassuring counter-narrative: a woman from rural Ireland allowing the country see itself again, in a more flattering light.

Forty years later, in 2018, Longford man John Connell was interviewed about his memoir *The Cow Book: A Story of Life on a Family Farm*, which, as he put it, is 'a story of modern rural Ireland and the wonderful people who inhabit it'. He would go on to be a regular contributor to RTÉ radio programmes. Bringing first-hand experiences of emigration, returning, family life on a farm, relationships, mental health and marathon-running, he captured the public imagination by connecting with a wider desire for a fulfilling life.

John Connell, writer and farmer, walking the land of his farm in Co. Longford. A familiar voice from his contributions to Radio 1.

Dick Warner records the tagging of a baby seal for the Radio 1 documentary *Naturepoint* on Lambay Island in 1978. From left: Dick Warner with John Flynn and Patrick Warner (his brother), both of the Wildlife Conservancy.

But what of the environment and natural world more generally being discussed on the radio?

From the archive I heard a discussion from Radio Éireann's weekly *City News Reel* with presenter John Ross in 1959 about Dublin Corporation's disposal of rubbish on Dublin's Sandymount Strand. The voice representing the residents made the case for the separation of materials into different types of waste for their disposal instead of landfill at a free natural amenity. The councillor representing Dublin Corporation argued that dumping and covering over was the most cost-efficient way to deal with the matter. *Plus ça change!*

It was 1987 before RTÉ dedicated a regular radio programme, *Outside Broadcast*, 'exclusively about the environment in this country', as its presenter Dick Warner described it – a green radio programme covering ecological matters in a way that was

Ella McSweeney with Mark Jessop on the Saltee Islands. Long associated with bringing us to the land and sea on her Radio 1 walking series *Shanks Mare*, as well as being a regular presenter of *Countrywide*.

accessible to listeners. The year 1987 was, after all, designated by the EEC as the Year of the Environment. Warner's definition of green was broad and included natural history, bird life, organic gardening, alternative energies and the activities of the Environment Awareness Bureau. Also covered were topical subjects such as acid rain, the Sellafield nuclear power station in England and the *Kowloon Bridge* oil tanker, which caused all sorts of environmental as well as economic catastrophes when it crashed against the coast of west Cork in December 1986. The programme aimed to provide information about networks and events across the EEC. Warner said listener involvement was key and that letters were welcome. And there on his first programme, too, was the music of early-morning birdsong. Joe Bollard, a contributor to the programme who was blind, said that hearing birds on the radio was for him like birdwatching, transmitted through the ears.

I first heard the *Dawn Chorus* on Radio 1. It was a Sunday afternoon, not long after moving into our house, and we were painting the walls of the upstairs front room. We had completed the messy but fun removal of the wallpaper over the previous weekend with a hired steamer – playing in our house, thrilled to have one after our flat, but where we couldn't do much to put our mark on it. We'd carried the paint back the day before on our bicycles, thinking we were so clever in choosing the smaller tins to balance them on the handlebars. The radio was on, getting our casual attention. I'd heard Gerald Barry sign off his *This Week* programme from the news studio, so I knew it must have been coming up to two o'clock and that we had better get on with the painting. The voice of presenter Derek Mooney caught us though, probably because it was a bit confusing: 'Well, good morning to you because it is morning,' he said, and continued: 'I know wherever you are it's just after two o'clock in the afternoon, but ignore your kitchen clock, because where we are it's just after 4.30 a.m.' (People still listened to radio programmes almost exclusively in real time and could tell the time from what was on the radio.)

Mooney said that with him were Jim Wilson of Cuskinny Marsh and his regular, knowledgeable sidekicks, Richard Collins and Éanna Ní Lamhna. And along with them, an array of technology: 'a rifle mic that looks like a giant sausage and if you're a keen football supporter, you'll have seen one of these things on the sideline on many an occasion. It's got a big fur coat on it, and it's very good at picking out specific sounds – you know when you hear the ball being kicked and it goes "theuu".' He explained that they were going to focus the mic to get the attention of a variety of bird sounds and try to identify them as the programme progressed and tell us a little bit about them. 'Is that right?' he asked the others with him. Yes, it was, they agreed.

Jim Wilson explained that Cuskinny Marsh, where they were in Co. Cork, was the meeting place of several different types of habitats – ideal to hear many different types of birds. Near the coast, a bit of woodland, next to a lake and arable farmland. Derek let listeners know it was still dark, a little bit cold and damp, but that they didn't mind, and in an upbeat voice quipped that they were doing it to secure our knowledge for the future.

Richard Collins explained that the sound was coming like a great wave across Europe, millions of birds waking up and singing, and had now reached Cork. It would show up in Iceland in an hour and a half's time and after that in Greenland and so on around the world as it had done for millions of years. Derek gave a shout to producer Michael Holmes for the time; Holmes said back that it was shortly after they had started and they all had a good laugh, knowing the time didn't really matter. At home in Dublin I unplugged the radio, quickly went upstairs and plugged it in again

Ornithologist Jim Wilson, presenter Derek Mooney and zoologist Dr Richard Collins at Cuskinny March for the broadcast *Dawn Chorus*.

and had the birds, Derek and his crew with us as we got on with our painting.

This was a modest first step towards what would become an international phenomenon involving radio stations across the world – including Russia and India state radios, sharing the sounds of the *Dawn Chorus* live from all over the world on RTÉ radio. Variations on it continue annually. In 2012 Derek Mooney remarked that 'The dawn chorus is one of the most magical experiences in nature: a multitude of birds of many different species all singing together in harmony as morning breaks and light begins to fill the skies. As our natural world's most impressive and renowned concert, it is almost as though it has been tailor-made for radio.'

Glen Hansard revealed how the sea area forecast inspired his Oscar-winning song 'Falling Slowly' when he was talking to Miriam O'Callaghan on Radio 1 in 2013. It is a sound deeply associated with radio for anyone involved in maritime pursuits, but also popular with the public – many lured by the exotic names of headlands and sea areas mentioned and by the reassuring, formulaic delivery of the bulletin. In 2001 meteorologist Gerald Fleming shared some insights on Colm Keane's *Radio Days* when talking about the story of the weather forecast and RTÉ Radio:

> When you are broadcasting a sea area forecast, the picture to have in your mind is of some fellow sitting at a bench [a workplace in a seacraft] about sixty miles offshore, going up and down in a five-metre swell and trying to listen to you on a crackly radio and write it down. So, you've got to read it at a slightly slower pace than you would normally broadcast, and you've got to be very clear. Of course, the formulaic nature of it is designed so that the information can be got across very efficiently.

He recalled reading an article in which people in Erris, Co. Mayo, felt others knew about them because their place got mentioned in the forecast, 'from Mizen Head to Slyne Head to Erris Head'. It had the ability to bring diverse people together, people scattered right across the island, and while RTÉ can get blamed sometimes for being too Dublin-centric, the sea area, as Fleming put it, 'takes us to the land of the waves, crashing on the rocks and the trawlers heading off for their day's fishing. [For] people living in very isolated, and at times lonely communities, whose lives are governed by the weather and the sea and by its various moods, the sea area forecasts touch those people's lives and brings them into mainstream.'

Weather reporting was possible again after the war officially ended on 2 September 1945. This telegram delivering a weather report for broadcast on radio Éireann illustrates that weather was back in the schedule again.

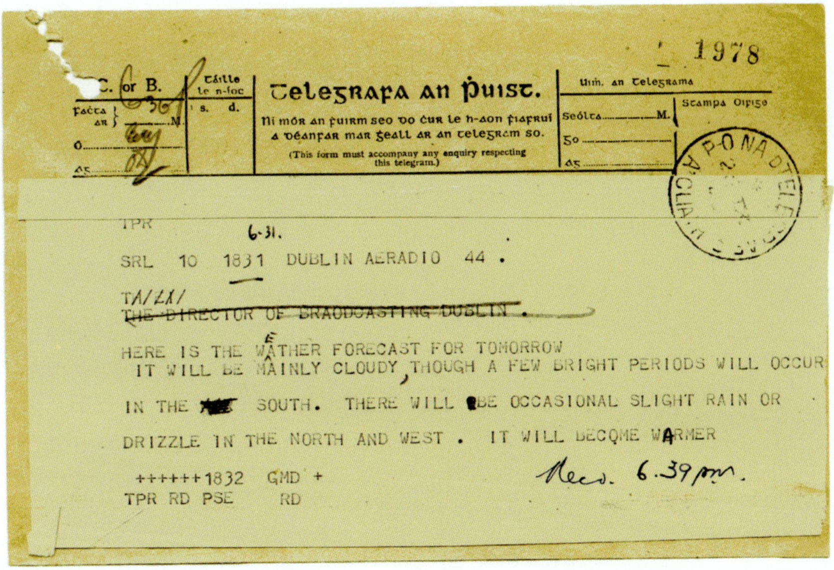
1978

C. or B. | Táille le n-íoc s. d.

Telegrafa an Ṗuist.

Ní mór an fuirm seo do ċur le h-aon fiafruí a déanfar mar ġeall ar an telegram so.

(This form must accompany any enquiry respecting this telegram.)

Uim. an Telegrama

Seólta ....M.

Stampa Oifige

TPR

6-31.

SRL 10 1831 DUBLIN AERADIO 44 .

TA/ZX/

~~THE DIRECTOR OF BRAODCASTING DUBLIN~~ .

HERE IS THE WATHER FORECAST FOR TOMORROW
IT WILL BE MAINLY CLOUDY, THOUGH A FEW BRIGHT PERIODS WILL OCCUR
IN THE [redacted] SOUTH. THERE WILL BE OCCASIONAL SLIGHT RAIN OR
DRIZZLE IN THE NORTH AND WEST . IT WILL BECOME WARMER

++++++1832 GMD + Recd. 6.39 pm.
TPR RD PSE RD

GLANMORE SONNET VII

BY SEAMUS HEANEY

Dogger, Rockall, Malin, Irish Sea:
Green, swift upsurges, North Atlantic flux
Conjured by that strong gale-warning voice,
Collapse into a sibilant penumbra.
Midnight and closedown. Sirens of the tundra,
Of eel-road, seal-road, keel-road, whale-road, raise
Their wind-compounded keen behind the baize
And drive the trawlers to the lee of Wicklow.
L'Étoile, Le Guillemot, La Belle Hélène
Nursed their bright names this morning in the bay
That toiled like mortar. It was marvellous
And actual, I said out loud, 'A haven,'
The word deepening, clearing, like the sky
Elsewhere on Minches, Cromarty, The Faroes.

The story of Ireland is inseparable from the story of its land and sea. From remote islands to family farms, to weather stations scanning the horizon, over the past century radio has allowed listeners to experience places they might never visit and lives very different from their own, while recognising something familiar in them all. These broadcasts form a rich audio portrait of this island nation: connected by water, shaped by landscape and continually making ourselves through sound.

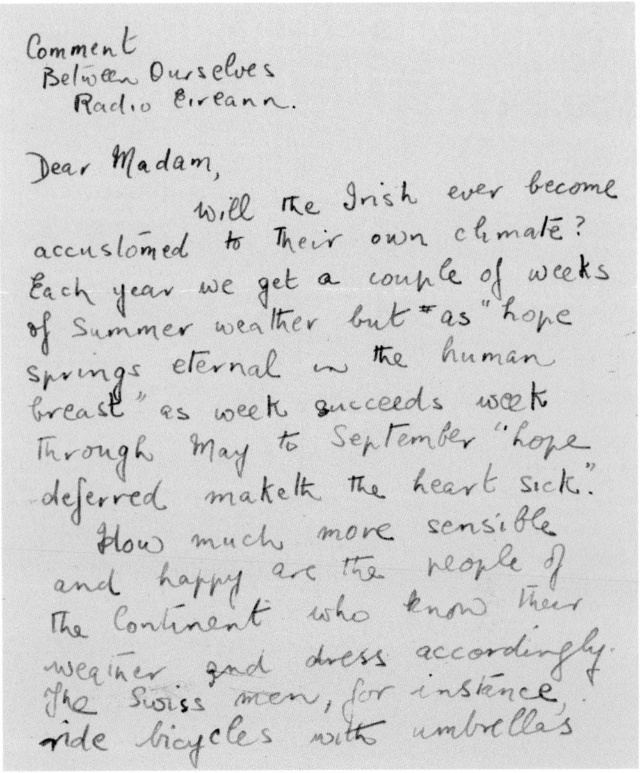

Comment
Between Ourselves
Radio Eireann.

Dear Madam,

Will the Irish ever become accustomed to their own climate? Each year we get a couple of weeks of Summer weather but *as "hope springs eternal in the human breast" as week succeeds week through May to September "hope deferred maketh the heart sick."

How much more sensible and happy are the people of the Continent who know their weather and dress accordingly. The Swiss men, for instance, ride bicycles with umbrellas

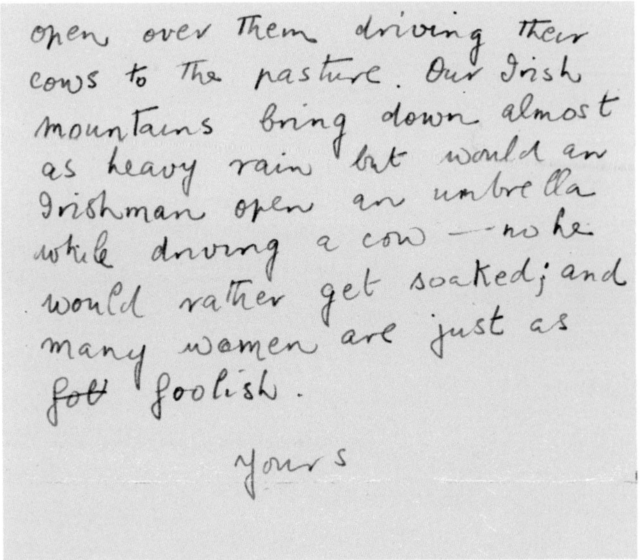

open over them driving their cows to the pasture. Our Irish mountains bring down almost as heavy rain but would an Irishman open an umbrella while driving a cow — no he would rather get soaked; and many women are just as foolish.

Yours

(Opposite page) ***Liveline*** **marked 100 years of BBC's on 2 January 2025. Seamus Heaney's 'Glanmore Sonnet VII' is also known as 'The Shipping Forecast'.**

Letter from a listener to the programme *Between Ourselves* in 1954 on the Irish relationship with weather.

COMMEMORATIONS
1916 AND THE BATTLE OF THE SOMME

Drawings from the Easter 1916 commemorative edition in 1966 of the *RTV Guide*.

The cover of the *RTV Guide*, Easter 1966.

Choosing which anniversaries are celebrated and how they are marked sheds light on where we are with their subject in time. And so, it has been with the 1916 Rising and the Battle of the Somme on RTÉ Radio.

For the fiftieth anniversary of the Rising the *RTÉ Guide* published an extensive commemorative edition. Radio coverage would have an emphasis 'on homage, on salutation, on the high emotion and the daring of that week, which not only aroused the moribund mind of Ireland, but afterwards fired that considerable part of the world which until then was sunk in Colonialism.' The defiant language contrasted with the radio silence in the same year on 1 July 1966, the fiftieth anniversary of the beginning of the Battle of the Somme.

Fast forward another fifty years and the contrast could not be more pronounced. Both the 1916 Rising and Battle of the Somme centenaries were accorded equal billing in the broadcaster's programming, part of a wider state commemoration in the Decade of Centenaries 2012–2022. *Reflecting the Rising* on Easter Monday 2016 was the largest historical and cultural state event, attended by an estimated 750,000 in Dublin. Amidst it was a day-long broadcast by RTÉ radio from the city centre.

RnaG broadcast the drivetime programme *Cormac ag a Cúig*, live from the Mansion House in Dublin on Easter Monday 2016.

The History Show hosted by Myles Dungan broadcast live from the Four Courts, with Evelyn O'Rourke reporting in from Liberty Hall, both significant buildings in the Rising. Many contributors to the programme referenced the impact of the Rising on civilians, from Katherine McSharry of the National Library reflecting on the suspension of bread and milk deliveries to Caitríona Crowe of the National Archives quipping that at least the fur coats lifted from Clery's department stores ended up in peoples' hands, because the premises was destroyed during the week anyway.

As well stories of its effect on citizens, the programme demonstrated links between the Rising and the First World War, with Julian Clancy reporting on how the medical experience of Dublin hospital staff from service at the Front assisted with treating gunshot wounds and difficult decisions around amputations.

In Sinéad Egan's report, Master of Holles Street Maternity Hospital, Dr Rhona Mahony reflected: 'The mothers of the fourteen births registered [for Easter Week 1916]

Joe Duffy presented *Liveline* from O'Connell Street on Easter Monday 2016.

Members of Donegal Youth Choir performing in Collins Barracks as part of the performance of A Nation's Voice on Easter Monday 2016.

would have come in through all the fighting and Mount Street was some of the worst fighting during the Rising'. A woman who set out from Capel Street had her fifth baby delivered en route to the hospital. As Dr O'Mahony said, 'she would have arrived with a baby in her arms and a bullet in her foot'.

On 1 July 2026 Radio 1 marked the centenary of the Battle of the Somme with short remembrances broadcast though the day; each preceded with the words: 'World War 1 on RTÉ Radio 1. Ireland's experience, remnants of the Somme.' During *Liveline* Cathy Hallesy described a sepia photograph of her grandfather's brother Fred Carter, an athletic who enjoyed his rugby. As an eighteen-year-old, along with two friends, he joined the ninth Division of the Enniskillen Fusiliers: 'I'm looking at the young face

Elizabeth II, accompanied here by An tUachtarán Mary McAleese in 2011, was the first British monarch to visit Ireland, and the Garden of Remembrance, since independence.

Irish and French dignitaries at the dedication of the France-Ireland Memorial, with its sculptor, Patrice Alexandre, third from left, at Glasnevin, 13 November 2016.

and the order to go over the top and I'm thinking of him, just facing into fire and having that split second of killing or be killed.' The *News at One* broadcast Sergeant Gerard White of the Irish Defence Forces reading from the last letter of Irish politician, poet and journalist Lieutenant Thomas Kettle of the Royal Dublin Fusiliers, which finished with the words: 'I am calm and happy but desperately anxious to live.' It was recorded at the Somme anniversary event that took place earlier in the day at Thiepval in northern France with Uachtarán Michael D. Higgins in attendance. On *Arena* presented by Sean Rocks that evening, Gerald Dawe read his poem 'Secrets'. Jennifer Johnston read from her First World War novel *How Many Miles to Babylon*? Eimear O'Connor talked about the involvement in the First World War of Irish artists William Orpen, Sir John Lavery, May Guinness and Beatrice Gubbins.

The two days were emblematic of a country giving itself time and permission to embrace a pluralist approach to historic centenaries. No longer was it a case of 'not knowing that we did not know' as Frank McGuinness put it on *Arena*. The story of the Irish in the First World War, had been silenced out of existence. A country growing up to acknowledge that the story it tells and how it is commemorated does not, and should not, stand still.

THE WIDER WORLD

CHAPTER 9

The familiar sight of Kish Lighthouse for emigrants who took the boat from Dublin over to Holyhead

Like so many in the 1980s, not long after graduating from university, I left the country. Some of us went to places where the Irish had, historically, gone in their thousands: Boston, New York, Chicago. London, Birmingham, Manchester, Liverpool. Canada and Australia. I headed for the other place. The less-worn route of the monks, the mercenaries, the wanderers, the wild geese. Au pairs, harvest hands, language tutors. I went to the Continent, to Spain. There, in Santander, I taught English for a living.

Out walking one morning I heard the music of 'Mná na hÉireann' in the air. I thought I was imagining it, but I wasn't. On taking the turn into the city's arched, stone-wrapped plaza, I spotted a musician. Perched in the best acoustical position for him to perform, curled in close to his harp, his head bent on what he was doing, as at home in himself and the music as if he were a local busker in Shop Street in Galway on a Saturday afternoon. After his set we got talking, me in my broken Spanish, him in his broken English, about Seán Ó Riada, Cantabrian and the Irish language, the sea; stories of migration and finding work, of places and people and what connects us across thousands of miles. Here I was not looking out from Ireland but back into it from outside, from another place and perspective, the North Atlantic coast of the Iberian Peninsula. What I saw and imagined from there are with me still.

Seán Óg Ó Ceallacháin presented Radio 1's Sunday-evening programme *GAA Sports Results* from 1953 until 8 May 2011. He entered the 2012 *Guinness Book of Records* as having the longest career in radio sports broadcasting. In 1948 he took over the Sunday GAA results programme from his father, Seán Ó Ceallacháin.

I like to think that what occurred earlier that day sparked what happened next, totally unexpectedly and out of the blue. Back in my flat, I paid more attention when tuning the dial on my radio, slowly moving it along until somehow it settled itself on a voice I immediately recognised: it was Seán Óg Ó Ceallacháin reporting on the weekend GAA results. Sometimes broadcasting really does allow us to exist in two places at once.

Today RTÉ is instantly available to us wherever we are. But it was a bizarre feeling back then to come across a live sound of home, somewhere else. During the late 1980s a newspaper sent from Ireland, say, could take a week to arrive. Phoning home was done rarely, and then only on Sundays, when it was cheaper. And of course, that distance from home was in some ways welcome. I loved living each ordinary day in Spain, without hankering after the place I had left, and yet sometimes being brought back positively to it too. The sometimes contradictory experience of living between places.

While I loved my life abroad, I was not alone in the joy I found in connecting with the sound of home, and nothing afforded the

Irish abroad greater pleasure than sports coverage – they would go to any extremes to catch a live broadcast of a match. Many were known to lean in with a transistor to a telegraph pole to catch even the slightest crackly coverage, particularly in Britain. By mid-century, the powers that be (or the enterprising Gus Ingoldsby) at Radio Éireann had realised that there was a large contingent of Irish missionaries in Africa who would relish a taste of the same. In 1953 he arranged with French national radio, in exchange for a modest fee, to relay live match commentary, which in turn would be transmitted to Brazzaville in the French Congo for re-broadcasting across Central Africa and beyond. How often I wondered why Micheál Ó Muircheartaigh would send 'greetings to our friends in Brazil' before I wondered in the writing of this book was he talking to the people of Brazzaville all along!

From the launch of Irish radio in 1926, there was an awareness that broadcasts might be heard beyond Ireland – by Irish people abroad and by non-Irish listeners – shaping the perception of both the Irish and Ireland. Dr Douglas Hyde's opening night speech on 2RN referenced this when he began:

> Ladies and gentlemen – I have been asked to open this broadcasting station tonight in the national language of Ireland. Our enterprise today marks the beginning not only of the New Year, but of a new era – an era in which our nation will take its place among the other nations of the world ... That much I have said in English for any strangers who may be listening in. Now I address my own countrymen ...

And the rest he would deliver as Gaeilge.

The same point was made in an *Irish Times* article on 1 January 1926 in advance of the first broadcast. In the article, the minister for posts and telegraphs, J. J. Walsh, acknowledged an audience beyond Ireland: 'the New Year marks the advent of Ireland among the broadcasting nations, and the bringing home to listeners-in abroad of the fact that this country has a separate culture and consciousness'.

Indeed, the first St Patrick's Day sheet for the Dublin Broadcasting Station 2RN in 1926 includes a 7.15 p.m. 'Director's Message and Greeting to all Listeners and Stations' with an indicated duration of one and a half minutes, showing that arrangements were in place for the St Patrick's Day programme to be relayed for broadcast by other stations, as they had been on the opening night of 2RN. In that message, director Séamus Clandillon said that he at least hoped that the station could connect with the Irish (and other people) outside the country who would be listening in on St Patrick's Day, of all days.

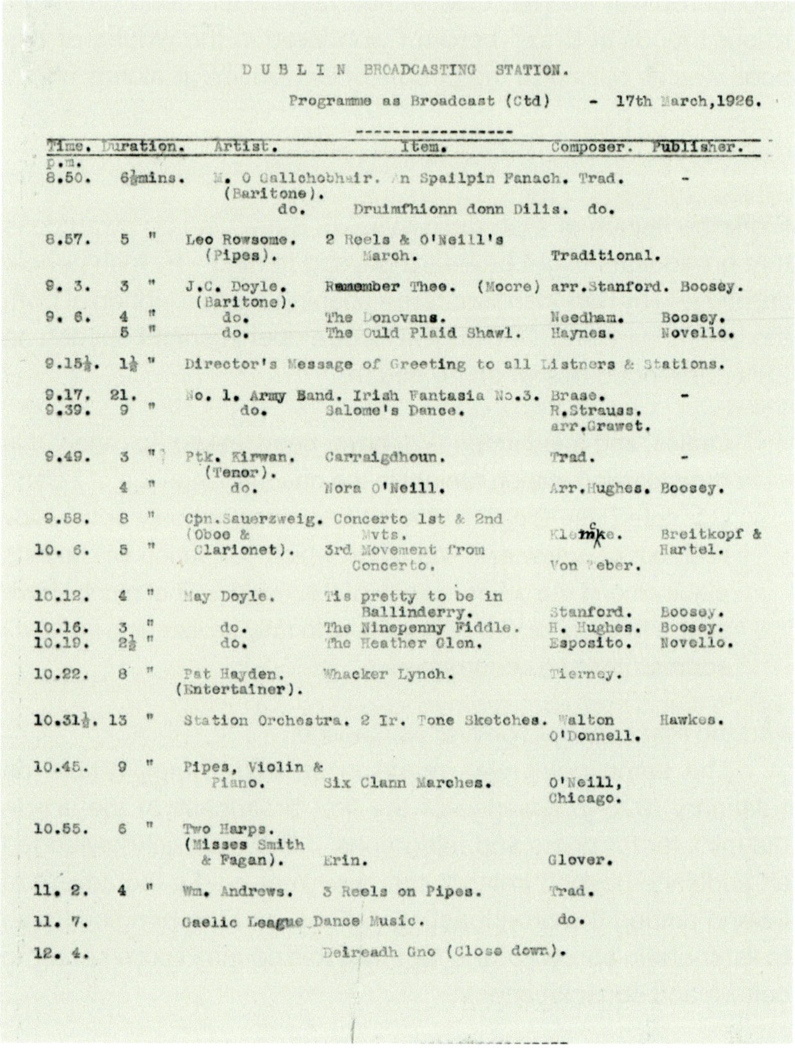

D U B L I N BROADCASTING STATION.

Programme as Broadcast (Ctd) - 17th March,1926.

Time.	Duration.	Artist.	Item.	Composer.	Publisher.
p.m. 8.50.	6½mins.	M. O Gallchobhair. (Baritone).	An Spailpin Fanach.	Trad.	-
		do.	Druimfhionn donn Dilis.	do.	
8.57.	5 "	Leo Rowsome. (Pipes).	2 Reels & O'Neill's March.	Traditional.	
9. 3.	3 "	J.C. Doyle. (Baritone).	Remember Thee. (Moore)	arr.Stanford.	Boosey.
9. 6.	4 "	do.	The Donovans.	Needham.	Boosey.
	5 "	do.	The Ould Plaid Shawl.	Haynes.	Novello.
9.15½.	1½ "	Director's Message of Greeting to all Listners & Stations.			
9.17.	21.	No. 1. Army Band.	Irish Fantasia No.3.	Brase.	-
9.39.	9 "	do.	Salome's Dance.	R.Strauss. arr.Grawet.	
9.49.	3 "	Ptk. Kirwan. (Tenor).	Carraigdhoun.	Trad.	-
	4 "	do.	Nora O'Neill.	Arr.Hughes.	Boosey.
9.58.	8 "	Cpn.Sauerzweig. (Oboe & Clarionet).	Concerto 1st & 2nd Mvts.	Klemke.	Breitkopf & Hartel.
10. 6.	5 "		3rd Movement from Concerto.	Von Weber.	
10.12.	4 "	May Doyle.	Tis pretty to be in Ballinderry.	Stanford.	Boosey.
10.16.	3 "	do.	The Ninepenny Fiddle.	H. Hughes.	Boosey.
10.19.	2½ "	do.	The Heather Glen.	Esposito.	Novello.
10.22.	8 "	Pat Hayden. (Entertainer).	Whacker Lynch.	Tierney.	
10.31½.	13 "	Station Orchestra.	2 Ir. Tone Sketches.	Walton O'Donnell.	Hawkes.
10.45.	9 "	Pipes, Violin & Piano.	Six Clann Marches.	O'Neill, Chicago.	
10.55.	6 "	Two Harps. (Misses Smith & Fagan).	Erin.	Glover.	
11. 2.	4 "	Wm. Andrews.	3 Reels on Pipes.	Trad.	
11. 7.		Gaelic League Dance Music.		do.	
12. 4.			Deireadh Gno (Close down).		

The Dublin broadcasting station 2RN's first St Patrick's Day programming on 17 March 1926. The 'Director's Message and Greeting to all Listeners and Stations' indicates that his message was relayed for live broadcast by other stations too.

2RN's director, Séamus Clandillon.

The *Evening Herald* of 8 September 1930 announced that, 'if the weather conditions permit', the speeches of President Hoover and Major Costes would be broadcast from Dublin, showing that experiments in long-distance cross-Atlantic broadcasting for Irish listeners were ongoing. From 1932, when Fianna Fáil got into power, Éamon de Valera, as president of the Executive Council, would speak on Radio Éireann to and about *na Gaelibh thar lear*, the Irish abroad.

He said that the new Athlone transmitter was a '*droichead idir na Gaeil in Éirinn agus Gaeil in imirce*' (a bridge between the Irish in Ireland and the emigrant Irish). In 1937 de Valera would use the radio to draw listener attention to talks about the new Irish constitution, ratified by the people on the same day as the country had re-elected Fianna Fáil. He spoke about the outstanding matters involving Britain, such as the ownership of Irish ports, tariffs and the partition of the country. His speech was broadcast to the League of Nations in 1937 and from the World Trade Fair in New York in 1939.

(Left)
Éamon de Valera shortly after his election as president of the Executive Council of the Irish Free State, on 9 March 1932.

(Right)
The Athlone transmitter in Westmeath, shortly after it was officially opened on 6 February 1933.

On St Patrick's Day 1943 Taoiseach de Valera's speech was broadcast on Radio Éireann and on an American radio station. The speech, extracted below, remains one of his most referenced:

> The ideal Ireland, that we would have, the Ireland that we dreamed of, would be the home of a people who valued material wealth only as a basis for right living, of a people who, satisfied with frugal comfort, devoted their leisure to the things of the spirit, a land whose countryside would be bright with cosy homesteads, whose fields and villages would be joyous with the sound of industry, with the romping of sturdy children, the contest of athletic youth, and the laughter of happy maidens whose firesides would be forums for the wisdom of serene old age. The home, in short, of a people living the life that God desires that men should live.

And of course, there was de Valera's speech on Radio Éireann as the war was nearing the end. It was a measured response expressing Ireland's right to its own decisions on neutrality and his disappointment with Churchill, who had three days earlier criticised what he saw as Ireland's unhelpful behaviour during the

NEW SHORT-WAVE STATION NEXT YEAR

The Radio Eireann short-wave transmitter, which will put Irish programmes on the air over the world, has yet to be purchased and erected, and is not expected to be in operation until next year. The new station will have a large staff.

Mr. Charles E. Kelly, Deputy Director of Radio Eireann for the past five years, has been appointed Deputy Director of the overseas programme. He is succeeded by Roibeard O Farachain, General Features Director.

Report in the *Irish Independent* of 31 August 1946 on the plans in train for Radio Éireann's short-wave transmitter expected to be in place in 1947.

war. The speech has often been said to be the moment when Irish broadcasting came of age. So that de Valera's speech would be heard more widely, Brian Durnin of Radio Éireann, a multi-linguist, broadcast it in French on the little-used short-wave transmitter, a symbolic rather than practical gesture since Radio Éireann was said to have no effective coverage in France. (Though it has been said that James Joyce, in Paris, was a daily listener to the station.)

Following Ireland's isolation during the war, in late 1945 plans were afoot for a short-wave transmitter which would allow for Irish radio to be heard all over the world. The same year, a weekly newsletter began to be broadcast from Paris and Rome. There was even the appointment of Charles Kelly as deputy director of the overseas programme and mention of extra staff being acquired. Writer Peadar O'Donnell, in his *Irish Press* column, suggested including voices of the Irish in London on Radio Éireann.

The likely uptake of general listeners in America to Radio Éireann was probably overestimated, due not only to the time

Peadar O'Donnell's Column

Own Folk on the Air

Some time ago, on a visit to a Wympie camp at Feltham, London, I suggested that the shows they stage there for their own entertainment could be put into shape to make a turn for a Radio Eireann broadcast.

I can think of few things more interesting to the countryside than to hear their own folk on the air from a hostel in Britain. London has so many good Irish artists round the film studios that any hostel could get help in this matter.

What hostel would like to make a start? I'd make life a misery for Radio Eireann chiefs until they agreed. (The truth is they jumped at the idea when it was whispered to them.)

RADIO EIREANN PROGRAMME

Sutton Coldfield,
Warwickshire.
August 8th, 1946.

Dear Sir—While sending you my half-yearly subscription for the "Meath Chronicle" let me, as a Meath exile, thank you for so promptly sending it to time each week. Indeed you may be interested to know it goes the round of at least five Irish families each week, not all of them from the Royal county.

One request we beseech you to grant us: Could you include Radio Eireann programme each week as we all "listen-in" to the motherland and it is so helpful to have a weekly survey? Thank you.—Yours

MARY BURT (nee Healy).

We shall do our very best to include the Radio programme in future. Ed. M.C.]

(Left)
Writer Peadar O'Donnell in his column of 19 September 1946 in *The Irish Press* suggests Irish voices in London be heard on Radio Éireann.

(Right)
Letter dated 1946 from Warwickshire, England to the *Meath Chronicle* requesting the inclusion of the Radio Éireann listings in the newspaper.

difference of the live broadcast but also to the short-wave broadcasting frequency not being in common use there. A letter from Warwickshire, England to the *Meath Chronicle* in Ireland dated 8 August 1946 requested that the newspaper include the listings for Radio Éireann because 'we all listen-in to the motherland'. Though the long-wave transmitter had been purchased, Fianna Fáil lost power in 1947 and the incoming government abandoned the enterprise to connect with international listeners. So, although the great new transmitter was never connected, the many improvements intended to bring Radio Éireann to a sufficient level for its launch on the world went ahead and began to give the station the semblance of a national broadcasting service.

The augmentation of the orchestra into the full Radio Éireann Symphony Orchestra from 1947 attracted international classical musicians and their families into Ireland. Though there

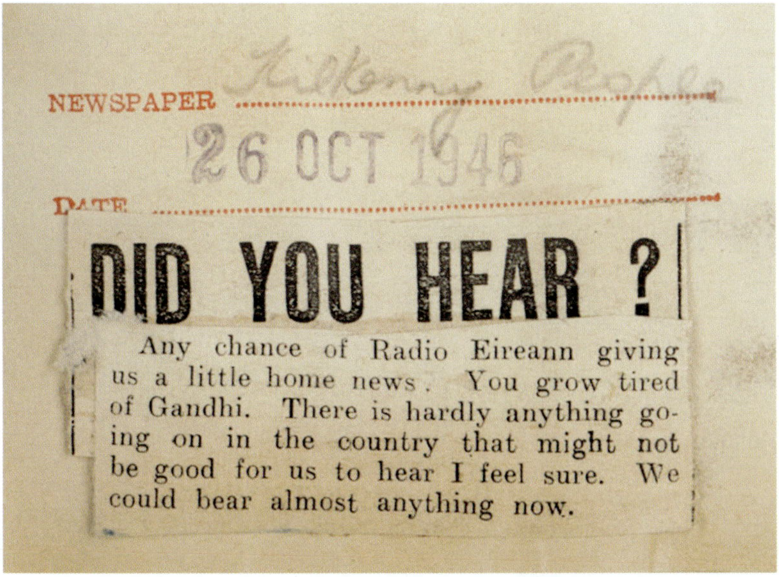

NEWSPAPER Kilkenny People
DATE 26 OCT 1946

DID YOU HEAR ?

Any chance of Radio Eireann giving us a little home news. You grow tired of Gandhi. There is hardly anything going on in the country that might not be good for us to hear I feel sure. We could bear almost anything now.

An article from the *Kilkenny People* in 1946 reflecting the perhaps sometimes random coverage of events outside the country in the period.

was a mixed welcome from Irish musicians, it brought a welcome expansion to Irish cultural life, considering the predominantly one-way outward movement of the population through the 1950s.

The station captured Ireland's engagement with world affairs, examples of which were membership of the United Nations and the country's role in international peacekeeping. *Congo Chronicles* featured in the radio schedule in 1961, and John Ross presented a fifteen-minute weekly programme on Monday evenings. Described as a report on the Irish troops in Congo, it included requests and personal messages from the soldiers for their families and friends at home. Captain Jack Miller would give the latest army news and stories.

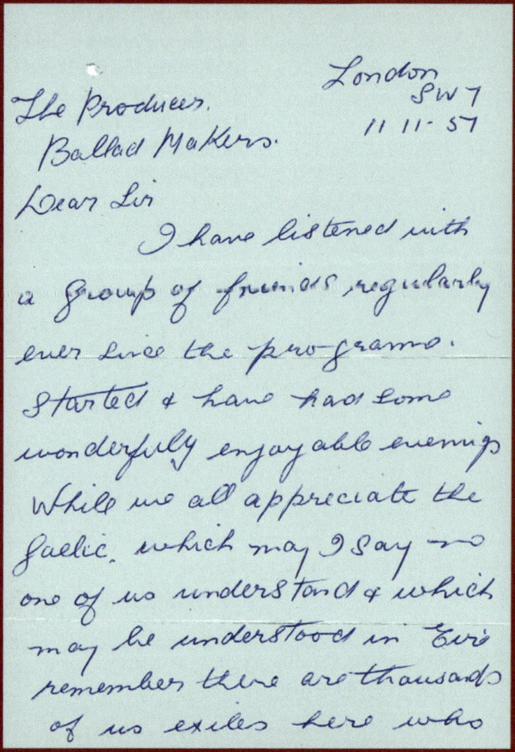

London
SW7
11 11-57

The Producer,
Ballad Makers.

Dear Sir

I have listened with a group of friends regularly ever since the programme. started & have had some wonderfully enjoyable evenings While we all appreciate the Gaelic, which may I say no one of us understand & which may be understood in Eire remember there are thousands of us exiles here who

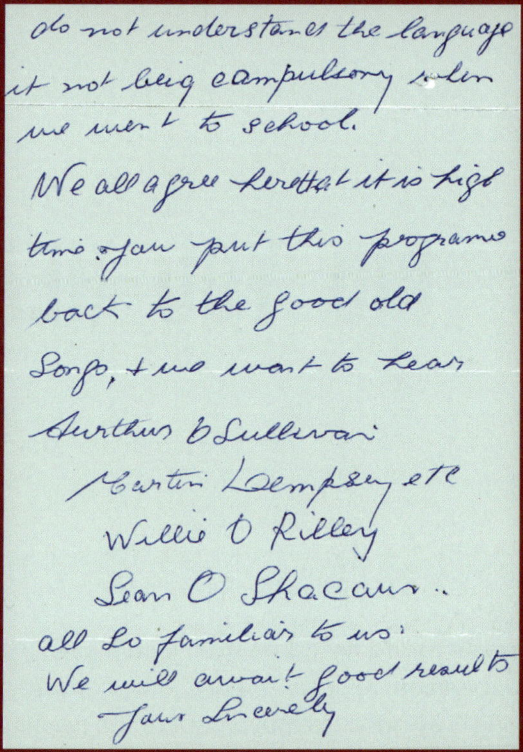

do not understand the language it not being compulsory when we went to school.

We all agree here that it is high time you put this programme back to the good old Songs, + we want to hear

Arthur O Sullivan
Martin Dempsey etc
Willie O Rilley
Sean O Shacan..

all so familiar to us:
We will await good results
Yours Sincerely

▲
A letter dated 1957 from an Irish listener in London stating that though he enjoys the Balladeers programme he doesn't understand many of the songs in Irish that are included because when he went to school Irish was not compulsory.

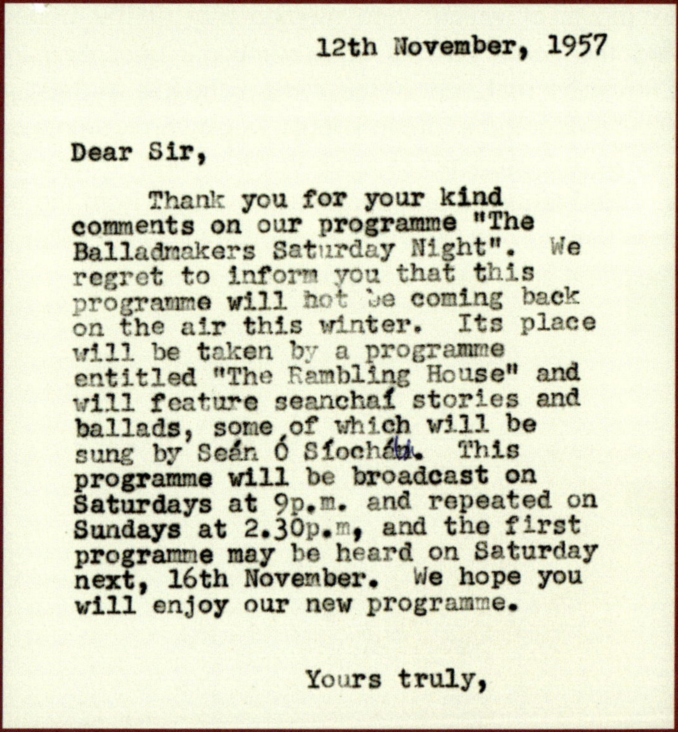

12th November, 1957

Dear Sir,

Thank you for your kind comments on our programme "The Balladmakers Saturday Night". We regret to inform you that this programme will not be coming back on the air this winter. Its place will be taken by a programme entitled "The Rambling House" and will feature seanchaí stories and ballads, some of which will be sung by Seán Ó Síocháin. This programme will be broadcast on Saturdays at 9p.m. and repeated on Sundays at 2.30p.m, and the first programme may be heard on Saturday next, 16th November. We hope you will enjoy our new programme.

Yours truly,

►
Response by Radio Éireann to the above letter.

Maureen O'Hara Home In "This Land Of Mine"

("Herald" Staff Reporter)

Maureen O'Hara, the film actress, one of whose films is appropriately called "This Land of Mine" and who refused about a year ago to acknowledge herself as a British subject when she applied for her American nationalisation papers, has brought home the nationalisation certificate, and hopes to show it to Mr. de Valera during her Dublin visit.

She arrived at Rineanna from the United States on Saturday, and is spending a three weeks' holiday with her parents at Churchtown Road, Milltown.

Miss O'Hara is very proud of the stand she took about her nationality.

"I think," she told me, "that I have created a precedent. It is, so far as I know, the first certificate of its kind that acknowledges an Irish nationality. It is certainly the first in California. There has been, I was told, a lot of trouble with Irish-born people about this very same thing."

Miss O'Hara feels that Mr. de Valera would be interested in seeing the actual document, and hopes that it will be possible for him to see her.

"BRITISH" ERASED

She showed me on her certificate where the word "British" had been erased, and on the success of her petition to the Courts the word "Irish" typed in its place.

A special note on the back of the document explains why the erasure was made.

Miss O'Hara's action pleased Mr. de Valera so much that he made a special reference to her in a speech he made in Waterford last January.

Said Mr. de Valera: "Miss O'Hara is a natural born citizen of this State and owes no allegiance to Britain."

Later in his speech he said: "Miss O'Hara was right when she asserted, as the Associated Press despatch states she did, that she owed no allegiance to Britain and therefore, had none which she could renounce."

Miss O'Hara whose parents are Mr. and Mrs. C. Fitzsimons had intended to make her Irish holiday a six weeks' one, but she has to hurry back in three weeks' time to make another film for RKO Radio called "Chicken for Sunday."

IRISH FILM

In the spring she is coming back to Ireland to make an Irish film based on Patrick Purcell's book, "The Quiet Man." It will be directed by a Galway man, John Ford and Victor MacLaglen will be in it.

"It has always been my ambition ever since I went to Hollywood to make an Irish film," said Miss O'Hara.

She is accompanied by her two-year-old daughter, Bronwyn Bridget Price.

Yesterday she visited her sister who is a Sister of Charity in a Dublin convent.

During her stay she will make a broadcast from Radio Eireann.

Evening Herald of 29 October 1946 report on the refusal of Irish actress Maureen O'Hara to declare she was British when applying for American nationalisation papers, and that she was shortly to broadcast from Radio Éireann.

Mrs Mary Ryan takes tea with her cousin President John F. Kennedy on the occasion of his visit to his ancestral home at Dunganstown, Co. Wexford on 27 June 1963.

One of the most glamorous come-good Irish emigration-slanted stories must surely be that of the election of John F. Kennedy as president of the US in 1961. So too, in its way, must have been the broadcast of his visit to Ireland in the summer of 1963. The visit was extensively covered on television but also on radio, as the daily broadcast sheets of those couple of days testify. Kennedy's Boston accent, especially when he pronounced Irish place names such as Dublin, Wexford, New Ross and Dunganstown, is still memorable today. His voice came to represent hope, progress and a more outward-looking Ireland. It reflected the international influences that began shaping the country in the 1960s and which would transform Irish radio by the end of the decade.

In contrast only months later, President Kennedy's funeral took place in Washington DC on 25 November 1963. RTÉ planned to broadcast the entire funeral event live with the use of Telestar satellite. President de Valera attended. Also, by special request of his widow Jacqueline Kennedy, a group of Irish cadets from the Curragh travelled to form final military honours at his graveside in Arlington Cemetery. President Kennedy had only recently been impressed by their drill at Arbour Hill in Dublin.

(Opposite page) RTÉ Radio schedule for 27 June 1963, that includes coverage of Kennedy's visit to Ireland.

REVISED PROGRAMME

RADIO EIREANN

ATHLONE 530 Metres 566 Kcs. DUBLIN & CORK 240 Metres 1250 Kcs.

THURSDAY, 27th JUNE, 1963

8.00 a.m. WEATHER AND NEWS
8.10 WEATHER FORECAST
8.15 MURRAY'S CAR RENTALS PROGRAMME
8.30 MUSGRAVE BROTHERS PROGRAMME
8.45 DUBLIN DAIRIES PROGRAMME
9.00 NEWS
9.10 WEATHER FORECAST
9.11 (approx.) MORNING MELODY
10.00 NEWS, WEATHER AND SIGNPOST
10.05 PRESIDENT KENNEDY'S WEXFORD JOURNEY:
Commentaries from New Ross, Dunganstown and Wexford Town as the American President visits the county of his ancestors.
Commentators: Conchubhair O Coileáin
Terry O'Sullivan
Brian Farrell
Seán Mac Réamoinn
Denis Meehan
John Bowman
John Skehan
Karl Jones
2.15 NEWS
2.30 JACOBS' PROGRAMME
2.45 BIRD'S ~~MAXWELL HOUSE COFFEE~~ PROGRAMME
3.00 NEWS HEADLINES
3.01 RECORDS
4.00 AN tUACHTARAN, EAMON DE VALERA receives President Kennedy at a garden party at Arus an Uachtaráin. Comment and interviews by Seán Mac Réamoinn and Brian Farrell
5.00 NEWS HEADLINES
5.01 COGAR, A LEANAI, le Peadar O Dubhda
5.15 AN DEALBH SPAINNEACH: Scéal le Deasún Breathnach. Pádraic O Néill a chóirigh agus a léirigh. Mír a 2 - Gaeil ón Spáinn (An dara craoladh)
Items/5.40 THE O'CONNELL SCHOOLS ACCORDION BAND conducted by Seán Kearney (RE recording) 27/6/63
5.55 WEATHER FORECAST
6.00 THE ANGELUS
6.01 NUACHT AGUS CAINT NA DEARDAOINE
6.20 ANNOUNCEMENTS: STOCK EXCHANGE PRICES: SIGNPOST
6.25 HERE'S THE DOCTOR: Weekly Health Talk
6.30 NEWS
6.45 THE KENNEDYS: A special "Meet the Clans in America" broadcast, compiled and presented by Micheál O hAodha, with Eoin O'Mahony as Chronicler, and including recordings of and interviews with
John Fitzgerald Kennedy, President of the United States
Robert Kennedy, Attorney General of the United States
Edward Kennedy, Senator for Massachusetts
Mrs. Robert Kennedy, and
Dr. Thomas J. Kiernan, Irish Ambassador to Washington (RE recording)
7.45 GUTH NA gCINNEDIEACH ADUAIDH: Fáilte roimh Uachtarán na Státaí Aontaithe arna cur in eagar ag Liam Mac Reachtain
8.00 NEWS HEADLINES
Items 8.01 STRING SONG: The music of the Audrey Park Ensemble introduced by Dick Cameron
8.30 BETWEEN OURSELVES: A magazine for women presented by Petronella O'Flanagan
9.00 NEWS HEADLINES
9.01 PRESIDENT KENNEDY'S WEXFORD JOURNEY: Edited recordings of this morning's commentaries

11.15 TUNE TIME with Tommy Dando and Frank Blowers
11.30 NEWS SUMMARY: LATE SPORTS RESULTS: WEATHER FORECAST: CLOSE DOWN

Irish emigrants leave on *Cunard* liner – scenes at Cobh, 3 March and 2 October 1953.

Michael O'Hehir was in America at the time and provided the commentary. Poor reception meant only twenty-five minutes of video was broadcast, during which President Kennedy's coffin was carried by horse-drawn caisson to St Mathew's cathedral in Washington for the funeral mass. The entire coverage went out on radio.

RTÉ Radio's weekly programme *World Report* began in 1969 to bring international news features to listeners. The idea, developed by RTÉ journalists Seán Duignan and Mike Burns, combined short current affairs reports and feature pieces from around the world, most of them presented by Irish contributors. The programme proved successful and continues to this day.

The establishment of RnaG in 1972 further expanded RTÉ's international reach. From its earliest years, the station built a global network of Irish-speakers, largely drawn from the Irish diaspora. This enabled it to report on and analyse major international developments in Irish, covering subjects ranging from British elections and the evolving European Union to political events in the United States and South Africa's first democratic elections.

Because many Irish speakers came from regions with long histories of emigration, the station recognised the importance of maintaining strong links with Irish-speaking communities abroad. Building and sustaining these two-way connections became a central part of RnaG's mission and identity.

A substantial series of half-hour Thomas Davis Lectures about emigration ran on RTÉ Radio in 1972, led by historian Joseph Lee. *Emigration and the Irish Experience* had eighteen radio talks from contributors in five countries. Lee remarked in the *RTÉ Guide* that, to discuss Irish emigration in the past 200 years was as much a discussion of Ireland itself. As he put it, 'there is scarcely a single political, social, economic or intellectual problem in modern Irish history which has not been influenced directly or indirectly by the experience of emigration and yet until the 1960s there was extraordinarily little research into the history of emigration'. Lee wanted to offer listeners an expansive look at the impact of the complex experience of the migrant on themselves, on where they left as well as the place to which they arrived. The series opened with a consideration of thoughts on Ireland and Irish society around emigration and went on to ask why little attention was given to a subject so central and common to the Irish and 'whether this shows up a loss of nerve, on the part of a stricken society, or are the causes more complex, and perhaps more sinister? Did we really want to stop emigration?'

A GREAT DAY FOR THE IRISH

THE Irish exiles in America suffer from a double dose of nostalgia. Looking back to the old country through their eyes is a new experience for which one is not prepared. (Life, one begins to realise, consists of going from one state of unpreparedness to another.) Fresh and bouncing from Ireland one did not expect to have to sigh over her or get involved in sad-sweet memories. But that is what happens when two or three Irish-born people are together.

The old times, smell of turf fires, driving cattle to a fair, eccentric parish priests, hard-whacking school masters—memory's cornucopia is spilt out on the carpet of the tenth floor apartment. "When I was a boy," "When I was a girl"—dances, devilment, drunks, broken-hearted mothers the Black and Tans, hard times, politics . . . . The hours speed by; the exiles are never done—wistful, wistful, wistful. Hurt Heart is not scheduled as a contagious disease and yet I caught it badly when in America.

I walked Ennis with a Clareman, street by street.

Stephen Rynne writes an introductory note to his Radio Eireann talk, "The Irish on Parade," St. Patrick's Day at 7.15.

We entered the churches and schools, leaned over bridges and took excursions out to Clarecastle and Spancelhill. A Clonakilty exile brought me on a conducted tour of his town. In Portland, Maine, I met an old fellow pottering around a convent garden; he was from Spiddal and we "did" Spiddal—almost acre by acre. They never shed the Old Land; it goes around with them, a built-in, irremovable part of their hearts.

I speak of the Irish-born exiles. The second generation of Irish-Americans are different. They are almost altogether stimulated by what 'Mother used to say long ago' and 'Father always held.' Somewhere or at some stage, a good share of sickly sentimentality crept into the handed down information. The second generation Irish are hard to take; an ordinary 1962 Irishman finds that he cannot live up to their exalted views.

The third generation are, when they think of Ireland at all, rabidly national. I met a college boy who led a revolt because a free day was not granted for St. Patrick's feast: he and his rebellious companions scrawled harps and shamrocks on the campus snow.

St. Patrick's Day is "The Great Day for the Irish"—all generations, any drops, or drop, of Irish blood. Long before that day comes, there are programmes and preparations, greeting cards and greenery. The casual Irishman out on a visit to the States is taken by surprise; his little hoard of humility (small island off Europe and all that) is blown sky high.

The great city of New York is turned upside down by parades, bands, banners and people going around wearing green ties and proud faces. 'There's Ireland for you!' The unuttered shout comes off Fifth Avenue each St. Patrick's Day. 'There's Ireland for you. What are you doing all these years, grumbling about rain, or rates, or political rackets?' Real, proud Ireland—seeing it shining out of those faces! Tipperarymen, Wicklowmen, Armaghmen, swinging by on a never-ending triumphant parade.

I think there are few things in the world so moving as the Irish exile's love of Ireland.

Stephen Rynne's introduction to his talk broadcast on Radio Éireann in 17 March 1962 about Irish-born exiles, as well as first- and second-generation Irish Americans.

The series discssed the causes and consequences of emigration and looked at experiences of those who went to Britain, America, Canada and Australia and included many examples of the Irish emigrant experience, including seasonal harvest work to places like Scotland, particularly from Donegal, as well as movement within Ireland itself.

Reflecting the intentions of the Thomas Davis Lectures, Joseph Lee's hope was that his series would contribute to a deeper understanding by listeners of Irish identity. The talks were broadcast the year Ireland joined the European Union (or the EEC, as it was called then) and give a glimpse into a country that would continue for some years to experience its own emigration before becoming a place into which others began to arrive in increasing numbers.

Detail of the French Embassy entrance and the building in Dublin in 1972 when the *Adminbrief* programme about the French Embassy on Dublin's Ailesbury Road at the was broadcast.

By 1975, thirty countries had accredited embassies in Ireland (today there are more than double that number). A weekly programme, *Adminbrief*, had none too catchy a title for what otherwise could be a subject with its own intrigue, glamour even. Producer Dick Warner told the *RTÉ Guide* that he intended to show what goes on in the embassies of other countries in Ireland. The French ambassador, His Excellency Count Pierre de Menthon, allowed the programme to record him and his staff as they went about their daily lives. France has had an embassy in Ireland since 1930, but in 1975 the nearest the average listener might get to France was by watching the Tour de France on television, a pilgrimage to Lourdes, or a modest camping holiday reached by ferry from Rosslare to Le Havre.

The 4 April 1975 edition of the *RTÉ Guide* featured a technicolour photograph of broadcaster Treasa Davison, then presenter of *Overseas and Cross-Channel Requests*, which were set up to keep people in touch with families and friends outside the country. She told the guide's John Walsh that she was inundated with requests and 'how seldom we have less than a hundred and fifty for any one programme [when] the most we can possibly fit is twenty-five'. There were estimated to be four million Irish in Britain and thirty million Irish Americans in the US.

Overseas and Cross-Channel Requests linked RTÉ Radio with radio stations around the globe that would transmit the programmes locally. Information posters were circulated in Irish clubs about broadcasts on local stations such as Birmingham's

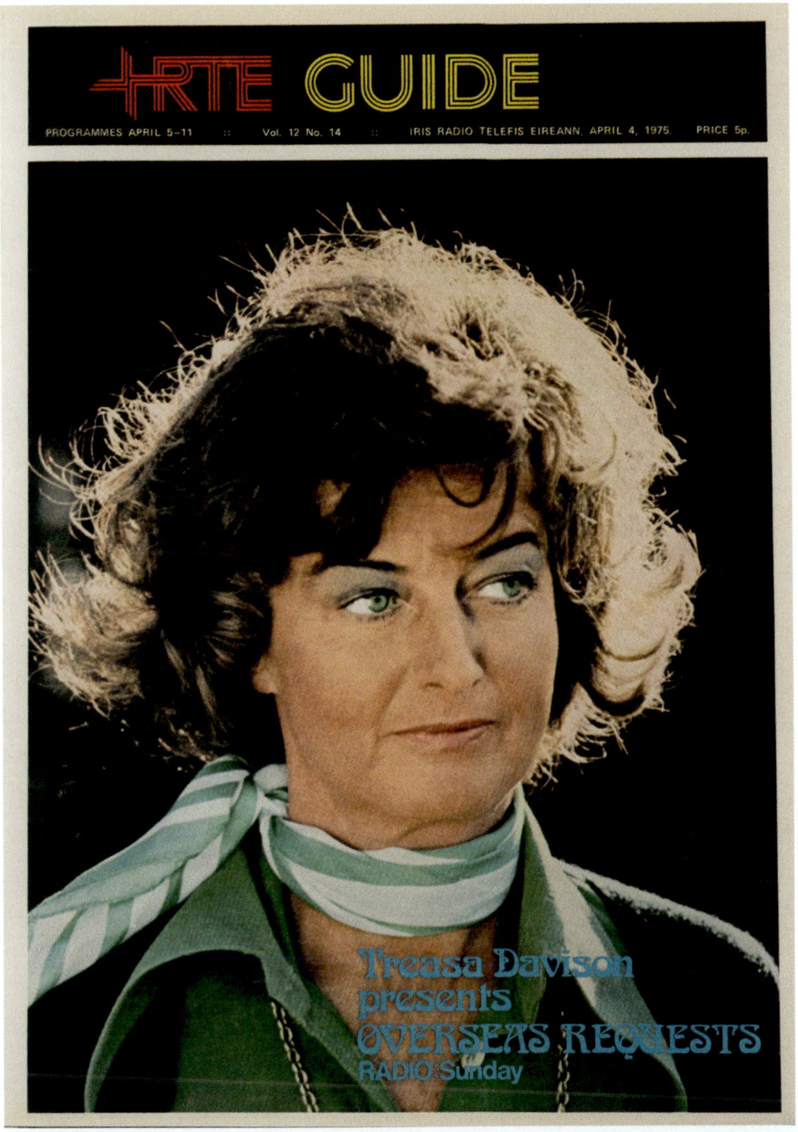

Presenter Treasa Davison on the cover of the *RTÉ Guide* of 4 April 1975. Inside she talks about her mission for her programmes 'to keep the Irish at home in touch with their friends and relatives abroad'.

BRMB as well as wherever RTÉ Radio was clearly available. The recent bombings in Birmingham hadn't helped producer Cathal MacCabe's task of 'extracting smooth-running links from the chaos of local stations around the globe', as he explained to the *RTÉ Guide*'s John Walsh. (Irish singer Brendan Shine was big in Bristol, from where an edition of the programmes was also broadcast.)

Overseas Requests was also associated with the UN Irish peacekeeping forces in Cyprus and Lebanon. Treasa Davison travelled there to record messages from the troops and relay pre-recorded messages from their families in Ireland. These were broadcast together on Radio 1 as 'Christmas Day Lebanon Requests'. Broadcasters Ronan Collins and Brenda Donohue would do the same in future years with relatives of families of UN Irish peacekeeping forces. During Christmas 1975 presenter Piaras Ó Gaora presented a programme from New York with Irish people there in a seasonal offering of RnaG, while playwright Criostóir Ó Floinn used the device of characters as fellow travellers on a train and plane to explore emigrant class distinctions in his new radio play *Cloch an Chairn*, produced by P. J. O'Connor.

By the 1990s President Mary Robinson had placed a light in the window of Áras an Uachtaráin as a symbol of welcome to the Irish diaspora. As the country held the presidency of the Council of Europe twice during the decade, familiarity with present-day continental Europe was growing. Paddy O'Gorman would become synonymous with his radio programmes – including *Queuing for a Living* and *Paddy's People*. O'Gorman was born in England to emigrant parents from Cobh, Co. Cork but moved back with his family to Ireland when still a child. For his 1998 series *A Future in Europe* he met young Irish emigrants in Germany, Holland, France and Spain. These countries were less popular: language, family, tradition and perhaps cultural similarities were such that emigrants do not usually think beyond England or its old colonies, even after years as members of the EEC. O'Gorman augured attitudes would change. This was not least due to the news of an undocumented Irish person being deported from the USA in handcuffs. He had made the error of going to Ireland for a family occasion, only to be arrested on his return to work in America – at least in Germany, being in the EU, he was legal! *A Future in Europe* related happy as well as hard-luck stories – the latter largely because of language barriers. One emigrant related how, in their opinion, 'along with the English, the Turks and the Yugoslavs, we Irish have become the skivvies of Europe'.

In 2000 RTÉ launched RTÉ Radio One World with broadcasts in Albanian, Arabic, Polish, Bosnian and other languages, aimed at

Ireland's growing international population. Cork was designated European Capital of Culture 2005 and former South African ambassador to Ireland Melanie Verwoerd presented a weekly celebration showcasing facets of Ireland's emerging multicultural population. Radio listeners learnt about others' experiences of the arts, music and sport.

Radio 1's *Sunday Miscellany* has been on the air since November 1968. It offers a weekly hour of new short writing read by the person who wrote them, interspersed with complementary music. In 2018, many of the essays broadcast to mark fifty years of the programme reflected on themes of emigration, immigration and new communities. Mary O'Malley wrote:

Kevin Barry reading 'An Exile in Liverpool' at Miscellany50 in the Project Arts Centre, Dublin in November 2018.

> I don't remember the late 1970s in Ireland, not for the usual reasons, but because I wasn't there. I had left by then and would stay away until 1986 ... Thousands were leaving as we came back. There was no welcome at Knock Airport, which was officially opened in 1986. The West was a wasteland. Many who left were illegal in America, getting in through Canada on holiday visas, doing the best they could. The human cost was stark – missed weddings, and missed funerals, missed christenings and celebrations. Couples were separated, houses repossessed, marriages fell apart.

Kevin Barry told about living in Liverpool in 1992:

> I was in romantic exile from the country where my stories occurred, but it was a close exile – on a clear night, you could almost shout across the Irish Sea, and the cousins would hear you. But it was an exile just about as far enough to put the place in perspective, to explain to me, as the sentences emerged, what it was I loved about Ireland, and what I hated, and I could feel that the stories were starting to work out for me on the page.

Melatu Uche Okorie wrote about 2006 and a woman called Dee who she met in her direct provision centre and who gave birth not long after arriving in Ireland:

> There was no one to call to share the news of my delivery. The local woman in my ward had people coming in and out to visit her, bringing her gifts and balloons. When it was time to leave the hospital, I was given a voucher for a bus. I carried my baby out and stood at the bus stop, a woman alone, inconspicuous to the people around me. I don't think anyone even noticed the tears running down my face.

Melatu Uche Okorie reading 'Why I Started to Write' at Miscellany50 in November 2018.

Experiences of emigration and of identity have increasingly found a place on the airwaves. *Continental Riffs*, the series that began in 2022 features artists, makers or producers, at least one of them with a connection to Ireland. They choose objects and experiences to prompt conversation. What began as a nod to the fiftieth anniversary of Ireland joining the EU has become a series of questions and ideas around belonging, identities, borders, intergenerational trauma, war, violence and home.

One such pairing was of artists Barbara Kneževic and Dragana Jurišic, who have known each other for nearly twenty years. Both are now Irish citizens. Dublin is their home. Barbara is part of a diaspora community, born in Australia to Yugoslav and Polish migrants. On her birth her father said that they were now Australian, so they would only speak English, cutting her off from the old languages of her parents – something familiar to Irish emigrants too. Dragana, from former Yugoslavia, gets confused about her native language because it has different names depending on how one reads history. The couple talk about their accented English and family histories. Barbara's grandparents went to Australia to escape the Second World War. Fifty years later Dragana Jurišic escaped to Ireland from her war-torn country. Barbara describes a black-and-white photograph of her grandmother, Aniela Putowska, taken for a German identity document during the war. On her lapel is sewn a patch with the letter P to signify that she was Polish: she was required to wear it while working as a forced labourer in Magdeburg in Germany. For Barbara the photograph represents the essence of Europe, the complexity of culture, of language, of history, of conflict; but it also symbolises hope, its persistence and survival.

One of the objects Dragana chose as a prompt to talk about her life was an enormous bible, which she salvaged from her family's burning house. She still doesn't understand why – she was running for her life, a sixteen-year-old girl with no interest in God or religion. The leather cover was so worn that it served as a pillow when she hid in basements during bombings. She found it again, having forgotten that she kept notes in the margins, like a diary, during the war. She hardly recognised the kid in the writing who thought that the trouble would stop, and that she would live happily ever after. Dragana says life is a grind. It's a beautiful grind. But it is a grind. For her the bible in many ways represents Europe, because so many things are wrapped in religion, in Christianity, which puts us in opposition with other parts of the world.

The Second World War-era photograph for the German identity card of Aniela Putowska, Barbara Kneževic's Polish grandmother.

In 1926 Douglas Hyde, on the opening night of 2RN, said, 'while not forgetting what is best in what other countries have to offer us, we derived from our Gaelic ancestors – from one of the oldest civilisations in Europe – the heritage of the O's and Macs who still make up the bulk of our country'.

Hyde was speaking for his own time in 1926. It would be for coming generations to make something of the country's recently-achieved independence, a matter that would often come to be played out and captured on radio during the century ahead.

SOURCES

The author gratefully acknowledges the assistance of the following institutions, archives, publishers and individuals for permissions, images, extracts, recordings and other materials reproduced in this volume.

ARCHIVES, LIBRARIES, CULTURAL INSTITUTIONS AND ORGANISATIONS

RTÉ and RTÉ Archives; Áras an Uachtaráin; National Library of Ireland; National Archives of Ireland; National Museum of Ireland; University College Dublin; University College Cork; Cnuasach Bhéaloideas Éireann / National Folklore Collection; UCD Archives; Trinity College Dublin; Dublin City Libraries; Sligo County Library Local Studies and Archives; National Gallery of Ireland; Fáilte Ireland Collection; Military Archive; GAA Museum at Croke Park; Abbey Theatre; Druid Theatre; ESB Archives; Irish Architectural Archive; Embassy of France; Irish Rugby Football Union (IRFU); Gaelic Athletic Association (GAA); DLR Heritage Office; Shankill ICA; Kilmacud Community Men's Shed; Ballinteer Active Retirement Group and Gael Linn.

PUBLISHERS AND PRESS

Faber and Faber; Mercier Press; New Island Books; Gill Books (Gill & Macmillan); Three Candles Press; Clonmore & Reynolds; Blackwater Press; Educational Company of Ireland; Kenny's of Galway.

NEWSPAPERS AND MEDIA

RTÉ Guide, *Irish Examiner* and Irish Examiner Archive; *The Irish Times*; Independent Newspapers (including the *Herald*); *Irish Press* and *Evening Press*; *Meath Chronicle; Kilkenny People*; *Hot Press*, Associated Press and Alamy.

WEBSITES

Dictionary of Irish Biography; Music Journal of Ireland; RTÉ and Century Ireland.

INDIVIDUALS
Caitríona Ní Mhurchú; Bairbre Ní Chaoimh; Caitríona Mc Laughlin; Christine Monk; Mark O'Brien; Mairéad Delaney; Joe O'Shaughnessy; Andrew Payne; Colum O'Riordan; Adam Staunton; Máire Ní Anluain; Cormac Kinsella; Hannah Keelin; Ronnie Murray; Ann Keegan; Máirín de Brún; Sandra Kelly; Deirdre Black; Jim Dunleavy; Paul McCarthy; Barbara Knežević; Cian McCormack; Barry McKevitt; Kathleen MacMahon; Alice Ryan; Dara McGee; Cormac Coyne; Damien O'Reilly; Derek Mooney; Kay Sheehy; Patrick Hugh Lynch; Jim Coughlan; Jim Kearney; Maighréad Ní Dhomhnaill; Paul Rouse; Ailbe van der Heide; Dean Kelly; Flor McCarthy; Mike McCormack; Tom Mc Sweeney; Claire Cunningham; Mary O'Malley; Geraldine Cryan; Barry Cunningham; Deirdre Black; Fergal O'Rourke; Lorelei Harris; Ella McSweeney; Ecaterina Grigorita; Malachy Moran; Gerry Scully, Damien O'Meara; Jan Ní Fhlanagáin; Linda Tobin; Zoe Reid; Treasa Mac Mánais; Alan Torney; Finola Doyle O'Neill; Rosemary Day; Eddie Bohan; Réamonn Ó Ciaráin; Christopher Morash; Michael Talty; James Conlon; Bridie Geraghty; Muireann Ní Chonaill; Conall Ó Móráin; Lynda Mulvin; Maeve Connolly; Sinéad Hogan; Sophie Byrne; Lisa Moran; Nathan O'Donnell; Annie Fletcher; Róisín Kennedy; Niamh O'Brien; Margaret Kelliher; Martina Devlin; Enda Wyley; Jane Leonard; Mary Sexton and Sarah Corcoran.

Special acknowledgement is also made for the use of materials relating to the *Model Housekeeping* publication (1930) and other images, photographs, artworks, texts and archival materials reproduced throughout this book.

IMAGE CREDITS

Permission to reproduce images on the page numbers listed kindly granted by the following institutions and individuals.

ARCHIVES, LIBRARIES, CULTURAL INSTITUTIONS AND ORGANISATIONS

RTÉ Archives: front cover, front inside Image 1, 2, 3, 4, 5, 8, 9,10, 11, 12 (left), 14, 15, 16, 17, 18, 19, 20, 22, 23, 24, 25, 26, 27, 28, 29, 30, 31, 45, 46, 47, 48, 50 (top), 51, 52, 53 (bottom), 56, 57 (top), 58, 65, 68, 70, 71, 72*, 73, 74, 75, 76, 77, 79, 80, 81, 85 (top right and left and bottom left), 86, 87, 89, 92, 96, 101, 104, 107, 111, 112, 113, 116, 121, 123, 125, 126, 128, 129, 130, 131, 132, 133, 134, 135, 136, 137, 139 (bottom left), 140, 145, 153, 154, 155, 159, 160, 161, 162, 163, 164, 165, 166, 167, 168, 169, 174, 175, 178, 179, 180, 181, 182, 183, 184, 191, 193, 194, 195, 196, 198, 199, 199, 200, 202, 204, 205, 206, 210, 220, 224, 225, 227, 228 (top), 229, 233, 235, 236, 238 (top), 241, 242, 244, 245, 246, 247 (left), 250, 253, 256, 257, 259, 260.

Photographed images from RTÉ Archives are by **Paul McCarthy.**

Alamy: 12 (right), 57 (bottom), 238 (bottom).
Blackwater Press: 214.
Cultural Relations Committee: 98, 127.
Digital Repository of Ireland: 120, 223 (top).
Educational Company of Ireland: 188.
ESB Archives: 150.
Fergal O'Rourke and IRFU: 34.
GAA Museum: 39, 41, 42.
Gael Linn: 190.
Irish Examiner Archives: 103, 118, 211, 217, 252, 254.
Maxwell Photography: 237.
Captain Jack Millar Private collection, Military Archives: 13.
Mercier: 90 (left), 100 (top) and right (bottom).
Mooney Goes Wild Team: 232.
National Archives (TAOIS/S3532/1): 122.
National Folklore Collection/Maighréad Ní Dhomhnaill (107), 211, 212.

National Gallery of Ireland: 207.
National Library of Ireland: 6, 7,10, 11,16, 36, 38, 44, 62, 64, 66, 95, 146, 147, 148, 149, 171, 172, 185, 186, 187, 189, 223 (bottom), 247 (right), 248, 249, 251, 255.
New Island: 90 (centre and right), 100 (bottom left and centre), 139 (bottom right).
Origin/RTÉ: 59 (top).
Sportsfile: 33, 49, 50 (bottom), 53 (top), 54, 55

INDIVIDUALS
Cormac Coyne: 209.
Claire Cunningham: 85 (bottom right).
Eamonn Doyle: 228 (bottom).
Brian Fay: 59 (bottom).
Eileen Heron: 225.
Ross Kavanagh: 82 (top).
Barbara Knesevic: 26.
Moira Lawson: 230.
Patrick Hugh Lynch: 239.
Treasa Mac Mánais: 139 (top).
Cian McCormack: 218.
Linda McGlynn: 169.
Máire Ní Anluain: 132 (bottom), 173.
Steve O'Connor: 202.
Joe O'Toole: 215.
Aisling Rogerson: 203.
Alan Torney: 141 (bottom).
Gerard Vickers: 82 (bottom).

All photographs on the endpapers are also from the above sources.

FURTHER READING

Balfe, B. *Radio Man: 40 Years of Broadcasting* (Gill & Macmillan, 2009)
Barry, K. and Smith, O. *Winter Papers* (Curlew Editions, 2001-present)
Boland, E. 'Domestic Violence', *American Poetry Review* 36:2 (2007)
Bowman, J. *Window and Mirror: RTÉ Television, 1961–2011* (Collins Press, 2011)
Bowman, J. *Ireland: The Autobiography* (Penguin Ireland, 2016)
Browne, T. *Ireland: A Social and Cultural History 1922–2002* (Harper, 2010)
Byrne, S. (ed.) *Dear Gay: Letters to* The Gay Byrne Show – *A Handwritten History of Ireland.* (Gill in association with RTÉ, 2023)
Coady, T. (ed.) *The RTÉ Yearbook* (Town House in association with RTÉ, 1989)
Cooney, John. *John Charles McQuaid: Ruler of Catholic Ireland* (O'Brien Press 1999)
Cronin, M. et al (eds) *The GAA: A Peoples' History* (Collins Press, 2009)
De Mórdha, D. *Raidió na Gaeltachta: Cultúr agus Pobal* (Cló Iar-Chonnacht, 2022)
Heaney, S. *Crediting Poetry* (Gallery Press, 1995)
Horgan, J. and Flynn, R. *Irish Media: A Critical Study* (Four Courts Press, 2017)
Gibbons, L. *Transformations in Irish Culture* (Cork University Press, 1996)
Gorham, M. *Forty Years of Irish Broadcasting* (Talbot Press, 1966)
Lee, JJ. *Ireland 1912–1985 Politics and Society* (Cambridge, 1989)
Maleney, Ian. *Minor Monuments* (Tramp Press, 2019)
McBride, O. and Gibney, J. *The Story of Us: Independent Ireland and the 1926 Census* (Irish Academic Press, 2026)
MacCarthy. F. *The Presidents' Letters: An Unexpected History of Ireland* (New Island, 2021)
McRedmond, L. *Written on the Wind: Personal Memories of Irish Radio 1926–76* (Gill & Macmillan in association with RTÉ 1976)
Morash, C. *A History of the Media in Ireland* (Cambridge University Press, 2010)
Ní Anluain, C. (ed.) *Reading the Future: Mike Murphy in Conversation with Twelve Irish Writers* (2000: Lilliput)
Ní Anluain, C. (ed.) *Miscellany50: Fifity Years of Sunday Miscellany* (New Island, 2019)
Ní Chinnéide, M. *Scéal Ghael-Linn* (Cló Iar-Chonnacht, 2013)
Ó Broin, L. *Just Like Yesterday: An Autobiography* (Gill & Macmillan, 1986)
O'Driscoll, D. *Stepping Stones: Interviews with Seamus Heaney* (Faber and Faber, 2008)
Ó Mórdha. S. *Seven Ages* (Séan Ó Mórdha/Araby Productions, 2000)
O'Toole, F. (ed.) *Modern Ireland in 100 Artworks* (RIA, 2016)
O'Toole, F. *We Don't Know Ourselves* (Head of Zeus, 2021)
Pine, R. *2RN and the Origins of Irish Radio* (Four Courts Press, 2002).
Rowley, E. (ed.) *More than Concrete Blocks Vol. 2: 1940–72* (Dublin City Council, 2018).
Watson, I. *Broadcasting in Irish: Minority Language Radio, Television and Identity* (Four Courts Press, 2003)

ACKNOWLEDGEMENTS

I am first and foremost indebted to those whose work during the first hundred years of RTÉ Radio meant there were programmes and references for me to work with to produce this book. And to those who listened to the programmes across that time, who made the work of producing them matter. It is a privilege to bring you this illustrated story of RTÉ Radio from my perspective.

I am most thankful to RTÉ Director of Audio Patricia Monahan who, in giving me the opportunity, support and time to deliver this contribution to RTÉ100, made all the difference; as well as to the RTÉ100 committee; Tara Campbell, Director of RTÉ Radio 1, for her invaluable encouragement and conversations, I am most appreciative; and to Elizabeth Laragy for her support of the accompanying radio and podcast series.

RTÉ Archives was central to the production of this book. I thank each of the staff for their always generous responses to my endless queries and questions. Only a fraction, of their knowledge and expertise they shared with me could, in the nature of a book like this, make it onto its pages. For my colleagues, too many to name individually, right across RTÉ, past and present, whose work inspired me to write this book, most especially for their generously shared knowledge, experience, encouragement and good spirit, that kept the energy that was required for the project going, I thank them.

To Mariel Deegan, Djinn von Noorden, Emer Brennan, Cassia Gaden Gilmartin, Des Doyle and Aoife K. Walsh, each of whom I will always be indebted to for their absolute belief in the book, their immeasurable encouragement, good humour and determination to make it what it is now.

Niall Scott, of Scott Tallon Walker, for his inspiring, detailed and generous conversations and walks, and for the Radio Centre building, a building whose every detail he drew before it ever physically existed.

Director Nora Hickey M'Sichili and Centre Culturel, Paris for the invaluable opportunity to spend time at the Cité de l'architecture et du patrimonie, Paris and with the Archives of Radio France, to consider public service broadcasting in a wider world context and for the time when this book began to take root.

Dr Paul Delaney of Trinity College Dublin's Department of English, literary consultant for the project, for his knowledge and specific reading of texts; Dr Orla Donnelly, who worked as a researcher; and Carlo Gébler for his conversations.

To my late father, Eoghan, who brought me and my sisters to RTÉ's 50 Years of Irish Radio exhibition in 1976, which prompted my curiosity about radio and how stories can be told. And for Brian and Nóra and Eoghan, always, and for everything.

INDEX

Clíodhna Ní Anluain is an award-winning radio producer, with RTÉ. She has reshaped some of RTÉ's most regarded programmes for contemporary listeners including *Sunday Miscellany*, RTÉ's Short Story Competition in memory of Francis MacManus, Davis Now Lectures and *Giant at My Shoulder*. She originated *SPOKEN STORIES*, a themed audio anthology of commissioned stories; *Continental Riffs*, conversations inspired by the continent of Europe; and the Decades of Centenaries series *Altering States: 100 years of the Arts in Ireland*.

Her arts practice has been supported by the Arts Council of Ireland and Le Centre Culturel Irlandais. Her film, *Pillar Talk* was part of Artworks: Remembering the Future at Visual, Carlow (2023). IMMA screened it and published her text in *Self-Determination: A Reader* (2023); Her solo exhibition 'Beautiful Things without [undue] Comment' was commissioned by The Dock, Co. Leitrim (2022). Her photographic work has been in the RHA Annual Exhibition. An award-winning commissioning theatre producer, museum curator and editor of anthologies for New Island, a collection of conversations for the Lilliput Press and talks for international architecture archive Pidgeon Digital.

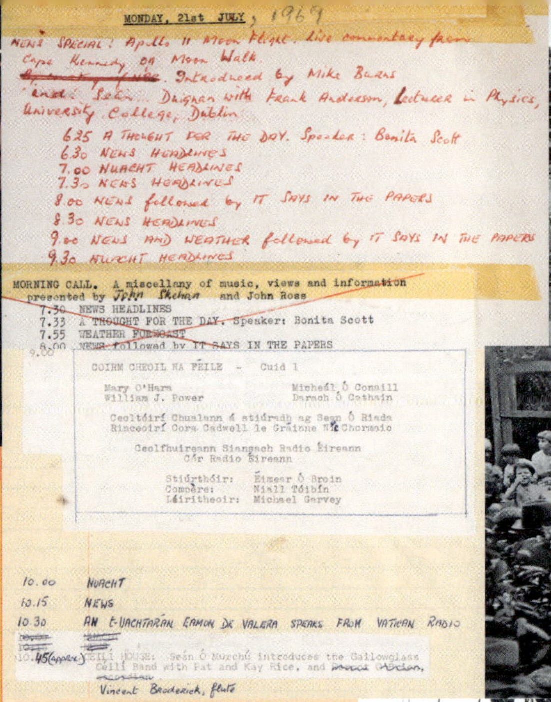

MONDAY, 21st JULY, 1969

NEWS SPECIAL: Apollo 11 Moon Flight. Live commentary from Cape Kennedy on Moon Walk. Introduced by Mike Burns and Seán Duignan with Frank Anderson, Lecturer in Physics, University College, Dublin

6.25 A THOUGHT FOR THE DAY. Speaker: Bonita Scott
6.30 NEWS HEADLINES
7.00 NUACHT HEADLINES
7.30 NEWS HEADLINES
8.00 NEWS followed by IT SAYS IN THE PAPERS
8.30 NEWS HEADLINES
9.00 NEWS AND WEATHER followed by IT SAYS IN THE PAPERS
9.30 NUACHT HEADLINES

MORNING CALL. A miscellany of music, views and information presented by John Sheehan and John Ross
7.30 NEWS HEADLINES
7.33 A THOUGHT FOR THE DAY. Speaker: Bonita Scott
7.55 WEATHER FORECAST
8.00 NEWS followed by IT SAYS IN THE PAPERS

9.00

COIRM CHEOIL NA FÉILE - Cuid 1

Mary O'Hara — Micheál Ó Conaill
William J. Power — Darach Ó Cathain

Ceoltóirí Chualann á stiúradh ag Seán Ó Riada
Rinceoirí Cora Cadwell le Gráinne Nic Chormaic

Ceolfhuireann Siansach Radio Éireann
Cór Radio Éireann

Stiúrthóir: Éimear Ó Broin
Compère: Niall Tóibín
Léiritheoir: Michael Garvey

10.00 NUACHT
10.15 NEWS
10.30 AN t-UACHTARÁN EAMON DE VALERA SPEAKS FROM VATICAN RADIO
10.45 (approx.) CÉILÍ HOUSE: Seán Ó Murchú introduces the Gallowglass Céilí Band with Pat and Kay Rice, and Vincent Broderick, flute

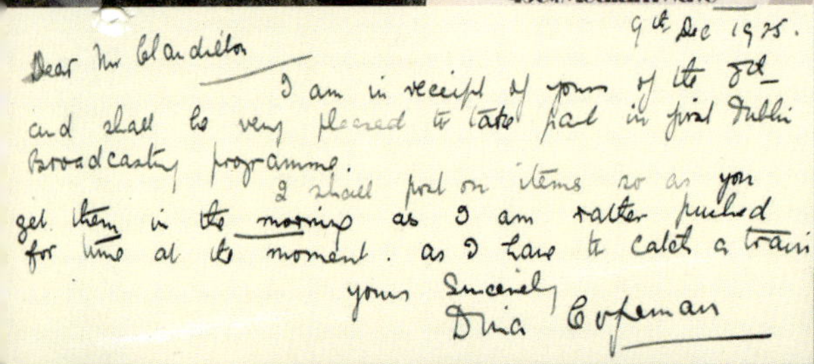

9th Dec 1925.

Dear Mr Clandillon

I am in receipt of yours of the 8th and shall be very pleased to take part in first Dublin Broadcasting programme.

I shall post on items so as you get them in the morning as I am rather pushed for time at the moment. as I have to catch a train

Yours Sincerely
Dina Copeman